D0700687

HONEYSUCKLE CREEK

ANDREW TINK is the author of celebrated books including *Lord Sydney: The life and times of Tommy Townshend*, *Air Disaster Canberra* and *Australia 1901–2001*. His biography *William Charles Wentworth* won The Nib award for literature in 2010. Before taking up writing, Andrew was shadow attorney-general and shadow leader of the House in the NSW Parliament, following an earlier career as a barrister. Andrew is currently an adjunct professor at Macquarie University.

HONEYSUCKLE CREEK

THE STORY OF TOM REID, A LITTLE DISH AND NEIL ARMSTRONG'S FIRST STEP

ANDREW TINK

NEWSOUTH

A NewSouth book

Published by
NewSouth Publishing
University of New South Wales Press Ltd
University of New South Wales
Sydney NSW 2052
AUSTRALIA
newsouthpublishing.com

First published 2018

10 9 8 7 6 5 4 3 2 1

NATIONAL LIBRARY OF AUSTRALIA
A catalogue record for this book is available from the National Library of Australia

ISBN: 9781742236087 (paperback)
ISBN: 9781742244297 (ebook)
ISBN: 9781742248721 (ePDF)

Design Josephine Pajor-Markus
Cover design Lisa White
Cover image Tracking Apollo 8 in lunar orbit on Christmas Eve, 1968. Photo by Hamish Lindsay.
Back cover image Photo of Tom Reid by Don Witten
Author photo Elizabeth Tink
Illustration opposite Based on photo by Ron Hicks
Printer Griffin Press

This book is printed on paper using fibre supplied from plantation or sustainably managed forests.

UNSW SYDNEY

The name Honeysuckle Creek and the excellence which is implied by that name will always be remembered and recorded in the annals of manned space flight.

Christopher Columbus Kraft Jr.
Director of Flight Operations, Apollo 11

Contents

Introduction

It was a Monday morning just after 6 am. Oblivious to the mid-winter chill that would normally have kept us in bed until at least 6.30 am, my family and I had been up for a while clustered around our kitchen radio. What we heard from almost 250 000 miles away was a human voice – an astronaut's voice – the voice of Buzz Aldrin.

> Six … forward … lights on … down two-and-a-half … forty feet … down two-and-a-half … kicking up some dust … thirty feet … two-and-a-half down … faint shadow … four forward … four forward … drifting to the right a little … OK … four forward … four forward. Drifting to the right a little. Twenty feet … down a half …[1]

As Aldrin's companion, Neil Armstrong, searched for a suitable place to land their lunar module, *Eagle*, Mission Control in Houston reminded them that they had only 30 seconds of fuel left. Would they make it? Or would they crash?

Then after what seemed like an eternity, we heard:

> ALDRIN: Contact light!

> ARMSTRONG: Shutdown.

ALDRIN: OK. Engine stop.

MISSION CONTROL: We copy you down, *Eagle*.

ARMSTRONG: Engine light is off ... Houston, Tranquility Base here. The *Eagle* has landed!

MISSION CONTROL: Roger, Tranquility. We copy you on the ground. You've got a bunch of guys about to turn blue. We're breathing again. Thanks a lot.[2]

According to our kitchen clock it had just gone quarter-past-six. In our different ways we were caught up in the high emotion of this moment: two men had landed on the Moon. Then, as my father was wont to do, we were hustled off to prepare for school. In my case it was a half-hour train ride from the Sydney suburb of Gordon to Town Hall, followed by a walk across Hyde Park to Sydney Grammar, an all-boys' school where I was in fourth form.

Having recently turned sixteen, I was interested in girls but awkward in their company. My schoolmates and I would clog up the entry ways of the old 'red rattler' carriages as our train snaked its way down the North Shore line. Standing nearby would be Monte girls who would get off at Milson's Point and SCEGGS girls going all the way to Darlinghurst. In our different groups we would talk at the tops of our voices, generally showing off. The girls would hitch up their skirts, just a little, while the boys would loosen their ties and effect a tousled hair look, something I could never quite pull off.

On any normal day the commuters in our carriage could not have presented a starker contrast, travelling in absolute silence and avoiding eye contact with those sitting beside them, even though they invariably travelled in the same or almost the same seats, day in and day out, year in and year out. Nearly every North Shore commuter read a paper, mostly the *Sydney Morning Herald*. In their cramped seats, the backs of which were uncomfortably stamped with the raised letters NSWGR, their only contact with each other was when they struggled to turn their broadsheets' pages. Without the need to utter a word, some commuters had learned the art of turning their respective pages in unison to minimise any disruption. The only words any of them ever spoke on these morning commutes were directed at us. 'If you continue misbehaving, I'll take your names and report you to the Headmaster; I'm an Old Boy you know!' Angry words like that.

But on the morning of Monday, 21 July 1969, the only morning I can ever remember it being like this, every commuter in my carriage was talking: to the person in front of them, to the person behind them, to the person opposite them or to the person beside them. And there was just one question on their lips. Exactly when would Neil Armstrong and Buzz Aldrin step out of the *Eagle* and begin their Moon walk, which NASA had promised would be televised live.

The *Herald* was reporting that the astronauts would rest for some hours before stepping outside, which in Sydney would coincide with everyone's evening train ride home. Some people said they would leave work early while others tried to figure out whether, if they stayed in the city, they could watch TV screens set up in department store windows. Still others, the ones dressed in the most expensive-looking suits, quietly discussed whether to allow the television sets in their executive dining rooms to be moved into their staff canteens. Complicating all these discussions was a quote in the *Herald* attributed to the Apollo 11 flight director, Clifford Charlesworth: 'We want to stick to the flight plan. But you know how flight plans are – sometimes you have to change them.'[3]

A few people in my carriage were able to listen to more up-to-date news than the *Herald* could provide, through earplugs attached to spindly cords that connected them to their transistor radios. One story which spread through our carriage like wildfire was a radio news flash that the astronauts' Moon walk might take place as early as 9 am. And when our train arrived at its first city station,

Wynyard, the commuter horde which alighted there scurried along the platform towards the exit signs with unusual swiftness.

By the time I arrived at Sydney Grammar, around 8.30 am, NASA's best estimate had been revised. It now seemed that Armstrong would emerge from the *Eagle* a little after 11 am. This news caused consternation among the teaching staff. They had assumed that the Moon walk would take place at the end of the day when only the sports coaches' after-hours training schedules would have been disrupted.

Alastair Mackerras, who had only recently been appointed Grammar's headmaster, was determined to see his school excel academically. For fourth form students like me, that meant getting a clean sweep of advanced passes in the externally marked School Certificate exams which were then just a couple of months off; in my case, 'straight As' in English, Science, Mathematics, History, French and Latin. There was a rigid daily program of classes or periods, each of which was commenced and ended by the ringing of a loud electronic bell.

The general sense among the Grammar staff was that disruptive as it might be to this strict academic schedule, the boys should be allowed to watch live TV of the Moon walk. And it appears that the headmaster, a classics scholar, reluctantly gave in. Every available TV set was requisitioned to the various classrooms; however, there were not enough sets to go around, so some missed out altogether. Others, including me, were assembled in the science block auditorium where a larger TV had been set up ready and waiting for the event.

However, the expected time came and went. On the screen in front of us Channel 9's Brian Bury did his best to fill in the unexpected programming gap by going over the *Eagle*'s landing earlier that morning, as well as interviewing guests such as Professor Stuart Butler from Sydney University's physics department. Although Butler was something of a celebrity, being well known to us all for his popular comic strip series, *Frontiers of Science*, even he started to wear thin after a while. Our teachers began fretting that we were wasting valuable class time. In many cases, including mine, classes resumed on a promise that when the Moon walk was about to begin we could return to the science auditorium to watch it. Just over an hour later the summons came and we found ourselves back in the auditorium, again watching Channel 9.

It was another false alarm as Brian Bury struggled again to fill the time. With our frustration growing, we started to play up as only pimply, cynical teenage boys can. Our voices, most broken but some not yet so, filled the auditorium with a cacophony of noise while a range of missiles, including tiny spit-balls launched from the hollow barrels of ballpoint pens, flew around. Then just as all of us were about to be placed on detention by our furious teachers, a split screen suddenly appeared behind Bury's head and we could see pictures of Houston's Mission Control coming to us in real time. With the prospect of a live cross to the Moon now only seconds away, the auditorium fell silent. According to the clock on the wall, it was just on 12.55 pm.

A few moments later an image came on the TV screen. But we couldn't make out what it was. To some it appeared to be upside down, to others back to front, and to all of us it seemed almost black for the most part with some strong light contrast at the bottom of the screen. As Brian Bury attempted to make sense of it, our restlessness started to return. Suddenly, the image on the TV screen seemed to adjust slightly. We could see a clearer picture, right side up and the right way around, of an unmistakably human form dressed in a spacesuit with what seemed like a giant backpack attached, stepping slowly down a ladder. The auditorium fell silent again. As Neil Armstrong stepped off the ladder's bottom rung onto one of the lunar module's footpads which was depressed just a couple of inches into the Moon's surface, we heard him say: 'The surface appears to be very, very fine grained, as you get close to it. It's almost like a powder. The ground mass is very fine … I'm going to step off the LM [lunar module] now …'

Armstrong then lifted his left foot backwards over the footpad's lip and dragged the tip of his boot in the lunar dust to test the Moon's surface. Apparently satisfied, he moved off the footpad and let go of the lunar module entirely, saying: 'That's one small step for [a] man, one giant leap for mankind.' Almost immediately our science auditorium was engulfed in a loud cheer.[4]

A few minutes later the image on our TV screen adjusted again to produce still clearer pictures, after which Buzz Aldrin made his way down the lunar module's ladder to join Armstrong on the Moon's surface. We watched as the astronauts attempted to plant a flagpole in the thin lunar soil, struggling because the pole was unbalanced by a strut holding out the stars and stripes. Shortly afterwards President Nixon appeared on a split screen from the Oval Office. Addressing the astronauts, the President said:

> Because of what you have done the heavens have become a part of man's world. And as you talk to us from the Sea of Tranquility, it inspires us to redouble our efforts to bring peace and tranquility to Earth. For one priceless moment, in the whole history of man, all the people of this Earth are truly one.[5]

With the astronauts' Moon walk scheduled to last for two-and-a-half hours, our teachers decided it was time for everyone to return to class. As the auditorium's double doors swung open and we emerged blinking into the sunlight, none of us doubted that we had just witnessed one of the greatest moments in history. For me this was reinforced when I arrived home for dinner. Staying with us was my grandfather who was nine by the time the Wright brothers first accomplished sustained powered flight in their heavier-than-air biplane at Kitty Hawk. Having survived over two years of hideous trench warfare on the Western Front during World War I and having struggled through the Great Depression, my grandfather was deeply moved that he had lived long enough to see a man on the Moon. He shook his head in disbelief as he recalled his boyhood in the small western Victorian town of Warracknabeal, where his family's only transport had been a horse and buggy.

The sense of elation I felt that day gradually faded and in 1972 I began studying law at the Australian National University. Towards the end of that year, I started dating a fellow law student, Marg Reid, who lived with her family in Canberra. We became close and were soon talking about spending time in Sydney together over the long summer break. This would not be possible, Marg said, until I had met her parents. Marg's mum was a lawyer, and her dad was the director of the Tidbinbilla deep space tracking station. We arranged to meet them early one evening when they would be having drinks with friends before going out to dinner.

As we approached Marg's front door, I could hear loud music – Isaac Hayes' *Shaft* I think it was – and lots of laughter. When the door opened, I saw a modishly dressed man with longish light brown hair who I judged to be in his mid-forties. He had a scotch in one hand and a cigarette in the other; and he was moving in time with the music. Marg introduced him to me as her dad, Tom, who immediately put down his cigarette, shook my hand firmly and in a pronounced Glaswegian accent greeted me warmly as 'Aundie'. I was then introduced to Marg's mother, Margaret, an equally strong but very different personality, and to their friends, John and Margaret Lodge.

Although a little awkward, the small talk between us all was friendly enough. Then Marg's younger brother, Nick, and sister, Danae, bounded into the room, loudly demanding to 'check out the new talent'. Everyone laughed heartily, Marg and I perhaps a little nervously as the colour rose in our cheeks. The Reids were fairly strict parents and they agreed to Marg visiting me in Sydney on condition that she stay with their friends, John and Chris Crowe in Mosman, rather than with me and my family in Gordon. Marg and I grew still closer during her short time in Sydney. Although we spoke mainly about our feelings for each other, we also discussed our families. And Marg told me that before running Tidbinbilla, her dad had been the director of the Honeysuckle Creek tracking station in the high country south of Canberra, when Neil Armstrong had first stepped onto the Moon. What neither of us fully appreciated back then was Honeysuckle's role as one of the three key Apollo tracking stations which had been placed equidistantly around the globe, including at Goldstone in California and near Madrid in Spain, to collectively provide continuous two-way communication between the Apollo 11 astronauts and Mission Control in Houston, Texas. Without the connections provided by these stations, the serried ranks of flight controllers in Houston would have been totally deaf, dumb and blind to the Apollo astronauts in space, just as those astronauts would likewise have been rendered completely invisible to their earth-bound controllers.

In a message to Australia's space trackers gathered in Canberra to mark the 40th anniversary of Apollo 11, Neil Armstrong said fiction writers such as HG Wells and Jules Verne had found ways to get people to the moon. But none of these writers:

> foresaw any possibility of the lunar explorers being able to communicate with Earth, transmit data, position information or transmit moving pictures of what they saw back to Earth. The authors foresaw my part of the adventure, but your part was beyond their comprehension.

It wasn't until I stayed with Marg and her parents over the New Year's weekend of 1972–73 that I began to realise the leading role Tom Reid had played tracking Apollo 11 and his extreme, almost pathological, reluctance to talk about it. Marg's family home was in Glasgow Street, Hughes, located high on the north-western slope of Red Hill. The back of the house faced out in the general direction of the Tidbinbilla tracking station, which was on the far side of the Murrumbidgee River and hidden from view by the Bullen Range. This positioning, as well as the street name which was the same name as Tom Reid's birthplace, may have been deliberate; or it may have been accidental. I never found out because as I quickly came to realise, Marg's dad disliked nothing more than talking about himself. I didn't know whether Tom's reticence was a Glasgow thing or whether it was more to do with his personality.

Besides I was far more interested in Marg. Sometimes my interest was directed elsewhere – I remember seeing a NASA Achievement Award in the Reids' living room. It was dated 20 October 1969, twelve weeks to the day in the United States, after Neil Armstrong had first set foot on the Moon. Addressed to 'Thomas Reid Manned Space Flight Station Director', this award read in part:

> For exceptional support to the success of the Apollo Program. His leadership, professional skill and personal dedication to the management and operation of the Manned Space Flight Network contributed significantly to the Apollo missions and greatly advanced man's capabilities to explore the vastness of extra-terrestrial space.

My mind flashed back three and a half years, to that chaotic morning in Sydney Grammar's science auditorium. 'Did you have something to do with the Moon landing?' I asked. Tom muttered something about 'team effort' and then changed the subject to English soccer which he keenly followed during summer while enjoying a whiskey or two. When I questioned Marg about this, she simply said that her dad didn't talk about work much and that it was best to let the subject drop. Marg clearly loved her father very much, just as he loved her. But Tom also had a strongly emphatic, almost stern side when his Glaswegian vowel sounds would noticeably harden and his naturally sharp hand and finger gestures that punctuated his everyday speech, would become sharper still. I could tell from the look on Marg's face that this sometimes unsettled her. It certainly unsettled me. Then again, I was dating the apple of his eye, and my sense was that Tom would occasionally put me just a wee bit off-balance to test my mettle.

By the time university resumed in 1973, it was clear to all that Marg and I were an item. Tom and Margaret included us in some of their regular dinner parties. I cannot recall any of Tom's tracking station colleagues ever being present and most of the guests were drawn from Margaret's circle: judges, politicians and diplomats.

With a dry, quirky sense of humour and always ready with a great line of banter, Tom was invariably the life of the party while Margaret was his prefect foil. Equally engaging and humorous in her own way, Margaret had a knack of disarming some of Tom's more risqué jokes with a perfectly timed look of mock horror. Sometimes I would be invited out to dinner with the family. Tom would always begin by ordering a bottle of Houghton's White Burgundy to accompany a relatively simple main course of steak or fish. On these semi-public outings, I noticed that Tom was much more circumspect. When a photographer snapped us all at a restaurant on campus he was less than impressed. It did not matter that the maître d' had allowed this fellow to click away at patrons in the hope that they might buy his pictures. Tom jealously guarded his privacy.

With the coming of spring, Marg and I would sometimes drive out to the Cotter River for a picnic. Then we would putter along in her little VW, all the way to Tharwa. The scenery was idyllic. The only reminder of civilisation was the Tidbinbilla deep space tracking station, a few miles down a side road. Once, in November 1973, we turned down this road and headed towards the Murrumbidgee River. Soon Tidbinbilla's massive 210-foot dish came into view. We knew from media reports that Tom's station was tracking NASA's unmanned Pioneer 10 spacecraft as it approached Jupiter. After passing this planet, Pioneer was to continue on into the outer solar system and beyond. Underscoring the potential of this voyage, the legendary cosmologist, Carl Sagan, had insisted that Pioneer be fitted with a gold-anodised aluminium plaque featuring the Earth's place in the solar system, images of a man and a woman and a stylised deep space tracking station dish, in case it was ever found by an intelligent being somewhere else in the Universe. NASA estimated that sometime during 2003, Pioneer would be more than 7 billion miles from Earth after which its signal would be too weak to track.[6]

By the time of our visit, Pioneer 10 had already travelled over 370 million miles. As the little spacecraft closed in on Jupiter,

Tom Reid and his team were tasked with uploading an electronic package of some 16000 commands to control its fly-past of that massive planet. Looking at the Tidbinbilla dish framed against a brilliant blue sky, I found the distances its signals had to travel difficult to comprehend. As if to put these distances into perspective, a couple of passenger jets on the Sydney to Melbourne run passed over head almost 6 miles above us. Apart from their jet engines' vapour trails, it was almost impossible to make out these aircraft with the naked eye. The only thing which gave away their precise position was the reflected light of the sun's rays coming off their fuselages. Had their pilots wanted to radio Canberra airport's control tower for any reason, communication would have been virtually instantaneous. By contrast, signals from Tidbinbilla to Pioneer took more than 40 minutes. As one paper put it, an error of just a few thousandths of an inch in the horizontal plane of Tidbinbilla's dish, would seriously affect its ability to communicate with the spacecraft. But all went well and close-up shots of Jupiter were played in real time, among them an image of its Great Red Spot that earned the Pioneer program an Emmy Award. Imagining how busy Tom would be on the day of our visit, Marg and I did not attempt to contact him. Instead, we walked around the station's public areas before continuing on to Tharwa.[7]

On weekends, I'd occasionally share a drink with Tom as he relaxed at home. His favourite tipple was Glenmorangie Scotch, a single malt highland whiskey. Despite his assurances that it slipped like silk over the tongue I wasn't tempted, preferring a beer instead. We had a common interest in rugby union and looked forward to the coming season. But I had difficulty sharing some of his other leisure interests such as watching TV shows relating to mathematics and physics. With astronomy it was slightly different because I remained keen to find out more about his job. To try to understand what Tom did, I read newspaper reports relating to his tracking station. And from these I knew that Tidbinbilla had been following Mariner 10. This robotic space probe had been launched by NASA in November 1973 and its unprecedented mission had been to fly by Venus using an interplanetary gravitational slingshot manoeuvre to bend its flight path so that it could then travel on to Mercury, the planet closest to the Sun. In doing so Mariner 10 had to endure solar radiation many times stronger than that experienced on Earth. Just what impact this might have on its communications was unknown. Nevertheless, on 16 March 1975 the probe was able to pass within 200 miles of Mercury's north pole, sending photos back to NASA via Tidbinbilla. According to the *Canberra Times* of 18 March, these images 'startled scientists with their clarity and detail'. Try as I might to get Tom to talk about this, he would deploy a slightly harder Glaswegian accent than normal to politely but firmly turn our conversation back to rugby.

Although Marg and I sometimes talked about marriage, it was always about other couples. At one stage she wrote to me about being frustrated by our stagnant relationship, now almost three years old, which could not be taken to its logical conclusion while we were at university. I felt the same. I simply couldn't see myself getting married until I had my law degree or, as my father put it, my meal ticket. I remained in college while Marg continued living with her family. By this time, we had been going out with other people as well as each other for almost a year. Towards the end of July 1975, Marg began dating a graduate law student. Before long, our mutual friends were warning me that it was serious. Sometime near the end of October there was a knock on my college room door. It was Marg. She'd come to see me at her parents' request, she said, before a public announcement was made. Then she told me that she and her new partner were engaged to be married at a ceremony which would take place at her Chapman home in just over three months. I felt as if I'd been struck by lightning. I gave Marg a hug and wished her well. Privately I was devastated.

In anticipation of what lay ahead, I walled myself in emotionally. I saw no more of Marg's family and did my best to avoid all contact with her. After finishing up at the Australian National University in 1976, I began working for a large Sydney law firm.

Despite having a job which was the envy of most of my old law school chums, I remained very unhappy throughout 1977 and the early part of 1978 until I heard on the grapevine around Easter that a girl I'd long had a secret crush on from afar had broken up with her fiancée. Her name was Kerry. I began to court her relentlessly. And after becoming engaged at the beginning of 1979 – I had to ask her twice – we were married that September. Among the many telegrams we received was one from Tom and Margaret Reid and another from Marg and her husband.

During the first half of 1981, Marg's mum was elected to replace a Canberra-based senator who had died unexpectedly. I wrote Margaret a congratulatory note and received a warm reply full of family news. Among other things I read that Marg and her husband had split up and that she had a new partner, Michael Phelps, one of Canberra's rising young commercial lawyers.

After over twenty years in the Senate including an historic six-year term as that chamber's first female president, Margaret Reid stepped back from active politics in early 2003. As a way of saying thank you to her many Sydney friends and supporters, Margaret hosted a luncheon in the NSW parliamentary dining room. Although Tom was elsewhere engaged, Marg Reid and Michael Phelps were there to represent the family. Marg and I sat next to each other and I asked her about her dad's reaction to a recently released film, *The Dish*. While claiming to cover Australia's Apollo 11 role, *The Dish* focused all the action on Parkes's radio telescope. In an Amazon customer review published in 2001 soon after the film's release on DVD, one of Apollo 11's most senior tracking engineers, Bill Wood, who had been at Goldstone on the day of the Moon landing, wrote:

> After viewing 'The Dish' one might think that Parkes was critical to the Moon landing mission. It was NOT! …
> The primary support was provided by the Honeysuckle Creek Manned Space Flight station located in Canberra. Honeysuckle Creek handled the telemetry, command and

> voice communications with all Apollo spacecraft ... If one does not take this film for more than Aussie entertainment, it makes for a good 101 minutes of DVD enjoyment. Just keep in mind that it is FICTION.[8]

Back then I was not aware of Bill Wood's review, but I clearly recalled the NASA Achievement Award Tom had received in October 1969 for his leading role in tracking Apollo 11. When I first saw the film's opening scene, with the 'Dish Master' played by Sam Neill walking towards the Parkes dish, I remember thinking, 'This isn't right: wrong place; wrong person'. It also struck me as very odd that *The Dish* hardly mentioned the Canberra tracking stations. When I queried Marg about this she told me that her dad had heard the film contained major factual errors and had refused to watch it. 'He should correct them,' I said. 'You know Dad,' Marg replied. 'He won't talk about his work.' And there the matter rested.[9]

After almost twenty years as a member of the New South Wales Parliament, I stepped down in 2007 due to ill health and took up writing. My first two books were biographies. At some point, I can't exactly remember when, Marg Reid asked me whether I'd consider writing a biography of her mum. But I begged off saying that I didn't want to rake over intimate details about my friends. In August 2010 some of my old law school chums organised a literary lunch for me at Canberra's Commonwealth Club which Marg attended. 'Mum would have been here too,' Marg said, 'but she's with Dad who's seriously ill.' About six weeks later Tom Reid passed away. Writing to Margaret to express my condolences, I said that if I ever got the chance I wanted to research and write about Tom's role in the Apollo 11 Moon landing.

That chance came in 2013 when I was commissioned to write a narrative history of twentieth-century Australia and was able to include a page or two about Apollo 11. Delving into Tom's role, I was astonished to discover that contrary to what had been inferred in *The Dish*, it was Reid's Honeysuckle Creek team that had transmitted to a worldwide audience of 600 million viewers, the live television of Neil Armstrong stepping onto the lunar surface. I was particularly influenced by a detailed critique called *The Truth about The Dish*, written shortly after the film's release by Mike Dinn, who had been Tom Reid's deputy at Honeysuckle on the day of the Moon landing. Among other things, Dinn said:

> The movie implies that Parkes was THE communication facility in Australia for Apollo 11 … The truth is that Honeysuckle Creek was … Parkes was and is a radio telescope – not a tracking station … Parkes had no transmitter and so could not send commands or voice to the spacecraft. So 'Parkes go for Command' as used in the movie is completely wrong and misleading … The movie studiously avoids stating that the first TV to Australia and the world came from Honeysuckle Creek.[10]

When I followed this up with Dinn, he pointed me to an audio tape and transcript of the moment NASA's man in charge of TV, Ed Tarkington, had chosen to broadcast Honeysuckle's signal. Left in no doubt that the footage of Armstrong's first step had come via Honeysuckle, I wrote up my two-page story. Having consulted Margaret and Marg Reid in the development of this tale we became more fully reacquainted. In late May 2015, Kerry and I attended Margaret's eightieth birthday party at the Commonwealth Club. There I caught up with many people I hadn't seen in forty years, as well as a few of Tom's former tracking station colleagues. I was also introduced to Tom's younger brother, Bob, who I had not met before. My meeting with Bob was significant because I had come to realise that Tom's role in the race to land a man on the Moon involved much more than simply supporting the Apollo 11 Moon walk. To do his story justice I would need to write a full biography.

The tracking stations at Honeysuckle, Goldstone and Madrid were as important to Apollo missions as the massive Saturn V rockets which launched their astronauts towards the Moon. Unless all three stations were in a state of 'Green Light Readiness', no Apollo launch could begin. While the key leadership positions at Goldstone and Madrid were held by Americans, there was not an American accent to be heard at Honeysuckle. It was Commonwealth government policy to staff Honeysuckle with Australian citizens or permanent residents. But it soon became apparent that Honeysuckle's first director was not up to the job. Honeysuckle's performance in NASA's simulated space tracking exercises, compared to the performance of the American-run stations at Goldstone and Madrid, was abysmal. That placed NASA's commitment to fulfil President Kennedy's promise to 'land a man on the Moon and return him safely to Earth before this decade is out' in jeopardy. With time running out and the Commonwealth government facing humiliation in the eyes of its American ally, it was Tom Reid who was called in as replacement director to whip Honeysuckle into shape. He achieved this with ruthless efficiency. Among other things, Reid replaced a number of engineers with technicians, some from the Navy, who although less academically qualified, had the drive, initiative and practical knowledge to work under extreme pressure in what was a cutting edge and at times experimental environment.

To write about how Tom Reid achieved this remarkable turnaround I realised that I would have to understand his life story, including his time in both the British and Australian Navies and his childhood in Glasgow. However, it wasn't until I met Bob Reid that I became confident I could chronicle Tom's formative years which were the source of his drive and focus.

Having told Marg Reid that I couldn't write biographies about people I'd once been close to, I began to rethink. Even from the little information I'd picked up at Margaret's party, I became convinced that Tom's extraordinary life story simply had to be told. 'Actual historical events should have been sacrosanct,' Dinn had written in his critique. 'But *The Dish* will perpetuate the myth that Parkes received the TV of Armstrong's first steps and transmitted them to the world.' As things stood it was as if Tom Reid and his Honeysuckle team had been written out of history. For me that was a wrong that had to be righted.[11]

Och, Tommy!

When Tommy Reid was ten years old, he told his mother, 'One day in my lifetime a man will walk on the Moon.' Although there wasn't much her precocious son said that surprised her, this particular prediction was too much. 'Och, Tommy!' Mary Reid replied. 'Yer bum's oot the winder and the rain's beatin doon on it!'[1]

Born in a small Glasgow tenement on 14 March 1927, Tommy Reid was the seventh in a line of first-born children called Thomas Reid. After completing an engine fitting apprenticeship with the Clydebank ship builder John Brown and Company in 1924, Tommy's father, Thomas the sixth, had found permanent work as a maintenance engineer with Fairy Dyes. As part of a team that supplied powdered dyes to Glasgow's textile manufacturers, he had been put in charge of the firm's automated filling and packaging plant. His wife, Mary, a down-to-earth, no-nonsense lady, worked for a gentlemen's tailor in Glasgow's premier shopping precinct, Sauchiehall Street. 'A dab hand with a needle,' according to her younger son, Bob, Mary's specialty was embroidered coats of arms.[2]

Glasgow had just over 1 000 000 inhabitants in 1927, over 600 000 of whom, according to a government report, had to make do with dwellings 'inferior to the minimum standard of the Board of Health'. Well over 40 000 families lived in one-roomed homes while a further 112 424 occupied places made up of a room and a kitchen. With decent accommodation at a premium, Thomas and Mary Reid rented a tiny Newlands Road tenement on Glasgow's tough south side. But about five months after Tommy's birth, his great-grandmother died and it was decided that Tommy and his parents would move in with his great-grandfather who occupied a relatively commodious tenement at 8 Crathie Drive in Glasgow's more genteel north. Located on the ground floor of a four-storey sandstone building, this tenement had two bedrooms, a large kitchen in which simple meals such as tripe and potatoes could be prepared, and a separate water closet, all accessed via a private corridor off the building's main entrance. However, there was no hot water on tap. For the family's weekly Friday night baths in a zinc tub, which when not in use was hung up on a hook behind a door, water had to be heated on a large coal-fired cast-iron range in the kitchen. Laundry was also a weekly affair and a single detached wash-house was shared by all the tenants. At the rear of the tenement building was a small grassed backyard.[3]

Tommy's seventy-three-year-old great-grandfather, Thomas the fourth, had been a brass finisher. Sometimes characterised as little more than repetitive polishing, this work was more demanding:

> The brass finisher requires more artistic taste than mechanical talent. He must have excellent development of Order, Size and Form … He needs also to be rapid in motion, and quick to see, and should have a fine and active temperament.[4]

Standing almost 6 feet tall and softly spoken, this Thomas did indeed have a fine and active temperament. With a keen interest in music, opera and reading, he was a bibliophile whose room was lined with books. A family photograph taken in mid-1927 shows the infant Tommy propped up on his great-grandfather's lap, with his father and grandfather standing proudly behind. The three adults epitomised what the author, George Blake, described as 'the Glasgow man':

> He is downright, unpolished, direct and immediate ... He can be terribly dangerous in revolt and terribly strong in defence of his own conception of order. He hates pretence, ceremonial, form – and is at the same time capable of the most abysmal sentimentality. He is grave – and one of the world's most devastating sardonic humourists.[5]

Already suffering from the early symptoms of prostate cancer that would eventually kill him, Thomas Reid the fourth was grateful for the love and support which his grandson's family gave him. He was especially fond of his great-grandson who, from his earliest days, displayed a keen intelligence. Cuddling the infant in his lap, Thomas the fourth would read to him for hour after hour, day after day, determined that this intellectually gifted tot would be given every opportunity to get ahead. Little wonder then that as soon as Tommy could talk, he was full of incessant questions. At the age of three he asked his astonished great-grandfather: 'Where does electricity come from?' This was at a time when there was no electricity in the Reids' tenement or anywhere else nearby. Although his family soon forgot the answer, they never forgot the question that passed into family legend.[6]

All the while the Reid family had continued to grow with the birth of June in 1929 and then Robert in 1933. This generational renewal gave a sense of enormous satisfaction to their great-grandfather whose cancer had spread, leaving him with only a few months to live. He was also close to his eight-year-old grand-daughter, Jan, who would catch a double-decker tram across town to enjoy his delicious home-baked muffins. Although Jan was Tommy's aunt, she was twenty years younger than her eldest brother, Thomas the sixth. And being about the same age as Tommy, they did much of their growing up together. But it was Tommy who stood out. His unquenchable thirst for knowledge became more apparent after he began attending Thornwood Primary School around the corner in Thornwood Avenue. The distinctive red sandstone school building was four storeys high, an intimidating fortress. What happened within that fortress and in other Glasgow schools between 9 am and 4 pm, has been described as follows:

> Classrooms, big or small, were much of a muchness everywhere, with blackboards, chalk dust, rows of desks in pairs, satchels at the feet and foreign bodies, various, stuffed down wally inkwells. The students stood in lines, did multiplication tables and regimented exercises to the teacher's counting, sat in rows, had Friday tests, gold stars and cleaning the board for good work … palmies for bad.[7]

While there was no mandated dress code, the boys wore ties and the girls wore pinafores. Judging from a class photo taken in 1937 when Tommy was ten, he and his classmates look happy and well fed notwithstanding that the Great Depression was then at its deepest. Staring intently at the camera, Tommy looks more alert, focused and intelligent than the others. With little interest in sport he might have risked appearing nerdy, except that he wasn't. Because even in those days there was something about him which suggested that he was more than able to look after himself and was someone not to be crossed.[8]

During the long summer holidays, Tommy's mother and grandmother would pack up their respective families and travel to Millport where, year in and year out, they rented the same large two-storey house across the road from the beach. One family lodged downstairs and the other on the floor above with the rent split evenly between them. Located on the Isle of Cumbrae near the Firth of Clyde, Millport was an adventure to reach: from Glasgow to Greenock by train, from Greenock to Largs by bus and from Largs to Millport by ferry. But the challenges of getting there were more than compensated for by the roomy accommodation, pristine beaches and sea air which together provided a welcome break from Glasgow's crowded tenement living. The Reid family's pleasures were simple. They avoided restaurants in favour of home-cooked fish and chips and sandwiches eaten on the beach. A treat was a walk into the village for ice cream and lemonade. In the early days Tommy and his aunt, Jan, would play on the beach, while in later years they rode bikes around the island. There was also an ancient cannon the provenance of which no one could remember that they enjoyed pretending to fire. On weekends their fathers would join them. Tommy's superbly muscled dad thought nothing of donning neck-to-knee bathers and taking Tommy for a swim. With an average June water temperature of just 11°C (52°F), a swim at Millport was not for the faint hearted. Most fathers settled for an ankle-deep wade in with their long pants rolled up above the knee and their shirt sleeves rolled up above the elbow. While most of them protected their heads with knotted handkerchiefs, Tommy's dad would emerge after a swim sporting a broad grin, his thick hair matted with salt water.[9]

Holidays down the Clyde Coast were an annual ritual for most Glaswegians who could afford them. But class distinctions, even among these more well-off vacationers, were clear to Jan and Tommy as they watched the latest arrivals disembarking from their steamers:

> There would be horse-drawn carriages for the posh people, and then a bit lower down the scale men with handcarts to take hampers and trunks to the boarding houses. And then there were the wee boys… 'Can I carry your bag, Mister?'

Tommy's favourite ships were 'puffers', so called because the smoke generated by their coal-fired boilers blasted out of their funnels. Crewed by a skipper, a mate and an engineer, these vessels carried a variety of cargoes including coal, building materials, sugar and paraffin.[10]

Tommy was fascinated by all things mechanical, among them a valve radio purchased by his father during the mid-1930s. Made out of dark wood in the form of an Art Deco-style cabinet and studded with buttons and dials, this radio was a piece of furniture in its own right. But the wet cell battery it relied on for its power required regular recharging. After a warning from his mother not to spill acid on his clothes, Tommy would be dispatched to the radio repair shop in nearby Dumbarton Road to top it up. Although Glasgow radio's 5SC broadcast a range of programs, the Reid family listened only to the news, perhaps to conserve the battery. For Tommy the radio itself was far more interesting than the latest BBC bulletin. Unlike most people who never asked about such things, Tommy wanted to know exactly how the battery worked and how it converted chemical energy into electrical energy; where the radio signals came from and how far they had travelled; how they were generated and how they appeared able to penetrate buildings; and what the valves did – preferably by taking them apart. However, any temptation Tommy might have had to unscrew the back of the cabinet and start fiddling around with the radio's innards was more than outweighed by what he knew would be his father's furious reaction. He turned to books for the answers he sought. What set him apart from his peers was his voracious appetite for reading. If you asked Tommy what he wanted for his birthday Jan Reid recalled that it was always a book. And his mother, Mary, was wont to say that Tommy always had his head in a book and a library in his head.[11]

By the time he was ready to leave primary school, Tommy Reid was devouring books on space. One was Jules Verne's *From the Earth to the Moon*, first published in 1865. It told the story of three men who were blasted into space aboard a projectile fired from a massive cannon located near Cape Canaveral. Another was its 1870 sequel, *Around the Moon*, in which the intrepid adventurers orbited the Moon and with the aid of auxiliary rockets, returned to Earth, where their landing was cushioned by deep water off the California coast. These stories were science fiction.

Later calculations by rocket scientists from Russia, Germany and the United States showed that human beings could not survive the shock of the massive explosion which would be needed to fire a manned projectile out of a cannon with sufficient force for it to reach its celestial target. Nevertheless, Verne's work excited the imaginations of these scientists: Konstantin Tsiolkovsky, Hermann Oberth and Robert Goddard. By their separate research efforts and the cross-pollination of their respective theories, a picture emerged of how a manned trip to the Moon might be undertaken. The work of Goddard, in particular, received worldwide press coverage during the 1920s and 1930s. And the idea of a multi-stage, liquid fuelled rocket began to take shape; one powerful enough to get to the Moon and back but one whose initial rate of acceleration would not kill its occupants. Goddard's work included plans for gyroscopic controls, gimbal-steering and power-driven fuel pumps, each of which would later prove vital to the development of space rockets. Although Robert Goddard was widely ridiculed by the press during the 1930s, Tommy Reid was transfixed by articles covering his theories and experiments.[12]

It was clear from Tommy's primary school results and his wider knowledge that he would perform well in scholarship examinations set by Glasgow's most prestigious secondary schools. On 31 August 1939, Tommy's parents were informed that he had won bursaries to Allan Glen's School, a selective school which had been founded in 1853, and to the Kelvinside Academy, an independent school established in 1878. But the very next day, as Tommy and his parents mulled over which of these schools to choose, Adolf Hitler invaded Poland. And the immediate futures of Tommy and his siblings were suddenly turned upside down.[13]

Keep Your Chin Up

Although the Spanish Civil War raged nearby, Guernica's market square was buzzing with bargain hunters on 26 April 1937. The scene was peaceful until a small force of German and Italian bombers appeared overhead. Without warning, they unleashed high explosive bombs and incendiaries, killing an estimated 2000 people. Outraged, Pablo Picasso immortalised this slaughter in a mural-sized oil painting. London had suffered from German air raids during World War I. But the ferocity of the attack on Guernica caused British strategists to recalculate. One estimate suggested that the Luftwaffe could carpet London with 100 000 tons of bombs in just two weeks, each ton capable of inflicting fifty casualties. As Adolf Hitler ramped up his war-mongering rhetoric the following year, plans were put in place to evacuate children from cities deemed to be at high risk.[1]

Having caved in to Hitler's demand for the annexation of German-speaking Czechoslovakia in September 1938, the British Prime Minister, Neville Chamberlain, was determined to make a stand when the Nazis invaded Poland a year later. In anticipation of a declaration of war, 120 000 children were made ready for evacuation from Glasgow whose shipyards were thought to be a priority target. Just 48 hours after weighing up offers from two of Glasgow's elite secondary schools, twelve-year-old Tommy Reid prepared for immediate departure, bound for a destination unknown. How long he and his siblings might be gone was also unknown; it could be weeks or it could be months. No one realised that it might be years. Accompanied by their parents, Tommy, his ten-year-old sister, June, and six-year-old brother, Bobby, made their way to Glasgow's Queen Street Station where the platforms were already thronged with thousands of youngsters kitted out the same way they were. 'Around your neck was an identification label, fastened by a loop of string,' Bob Reid later recalled. 'You were also given a gas mask in a cardboard box and a can of beans in a brown paper bag.' Soon after their arrival at the station, a marshal told Thomas and Mary Reid that their children's destination was the village of Kirkmichael, a few miles south of Balmoral Castle. To get there took 4 hours on the train to Perth, followed by another hour on a bus. Not knowing when she would next be able to see her children, Mary became visibly upset. Hiding his angst with a stiff upper lip, Thomas told each child in turn: 'Keep your chin up.'[2]

By the time the Reid children reached their final destination, a country house with a well-appointed drawing room, it was past midnight. One of the few things Bob Reid remembers about the place is a bowl heaped with sugar cubes. Having never seen a sugar cube before, Tommy, June and Bobby pocketed some. And after lights out, they happily chewed through their stash. The first reports Thomas and Mary Reid received about their children were positive: they were physically safe; they were being billeted in a comfortable home; and they had not been split up. However, there was a problem. Swamped with evacuees ranging from preschoolers to senior high school students, the local education authorities couldn't cope. Tommy the double bursary winner, found himself spending his school days with children as young as five. Infuriated, his father became a formidable advocate in his elder son's cause, resorting to threats of legal action against the local authorities. This had the desired effect and the three children were relocated 13 miles south to the larger town of Blairgowrie where Tommy could attend high school.[3]

Although not as intellectually gifted as his elder son, Thomas Reid was nevertheless as sharp as a tack and ambitious in his own way. In his younger days, he had been a keen athlete and a member of the Maryhill Harriers, competing against Eric Liddell whose gold medal performance in the men's 400 metres at the 1924 Olympics was later immortalised in the film *Chariots of Fire*. Thomas was also a fine golfer who kept up with the latest fashion, purchasing a pair of plus fours before they became popular on the links. Curious about what his neighbours might think of him in these newfangled loose-fitting pants which were tied off just below the knee, he walked down the street in them. And when no one laughed, they became part of his regular weekend attire.[4]

As his children settled into Blairgowrie and Britain moved to a total-war footing, Thomas Reid was directed to leave his job at Fairy Dyes. His engineering skills were required back at John Brown's shipyard where the battleship, HMS *Duke of York*, was under construction. With a keel almost 750 feet long and armour plating up to 15 inches thick in places, this behemoth was the pride of the yard. A member of its so-called black squad, Thomas was on call for any tricky engineering jobs where extra support was required. At thirty-six, he would have been at the upper end of the range for military call-up. But being in a reserved occupation, his uniformed duties were restricted to acting as an air raid warden. These responsibilities together with the general restrictions now imposed on all but essential travel, meant that Thomas and Mary had few opportunities to visit their children. Still, they managed to keep in touch. And what filtered back to Thomas about his elder son's schooling continued to disturb him. The high school Tommy was attending did not provide a complete secondary syllabus. There was only one school in the Central District of Perthshire where students could study subjects such as higher-grade mathematics, physics and chemistry for the Scottish Education Department's Senior Leaving Certificate: Morrison's Academy in Crieff. As Thomas saw it, his elder son had won bursaries to Allan Glen's school and to the Kelvinside Academy, both of which offered these subjects. But when war was declared these bursaries had become worthless. The Education Department, he argued, was therefore obliged to arrange a place for Tommy at Morrison's Academy and to convert his now useless bursaries into a Morrison's bursary. Although founded by a private endowment in 1860, Morrison's received substantial public funding which gave the Education Department some say in admissions. Overwhelmed by the large number of evacuees from all over Britain, the department was unresponsive at first. Thomas threatened legal action for a second time, triggering a formal arbitration. As a result, Tommy was awarded a bursary to Morrison's. After months of upheaval, he was finally able to knuckle down to the five years of secondary school which lay ahead.[5]

Because of the influx of evacuees, Morrison's was full to overflowing: the swimming pool had been floored over to provide extra accommodation; the memorial hall was being used as a classroom; and even some rooms in the janitor's house were pressed into service. With all boarding places already taken up, Tommy would have to attend Morrison's as a day boy. But Blairgowrie was too far away. So the Reid children were moved again, this time to the village of Braco, about 10 miles south of Crieff. This allowed Tommy to commute to Morrison's by bus while June and Bobby went to the local village school.[6]

During their eighteen months at Braco, the Reid children stayed in Ardoch House, a massive three-storey pile crowned by a prominent pitched roof and four large chimney stacks. Built in 1795, it had been requisitioned at the beginning of the war to house up to forty evacuees in dormitory style accommodation. A physical education teacher, Miss Crawford, was in charge, supported by an all-female staff. When they were not at school, in church or picking potatoes to support the war effort, the evacuees were free to roam around Braco village during the day. Soldiers often camped in the surrounding countryside while conducting training exercises. During these manoeuvres, the village would be defended by a red team, while a blue team sought to flush them out.

Sometimes June and Bobby would point out to a blue team soldier where a red team member was lurking, whereupon a furious military umpire would chase them away. Such mischief making was of no interest to Tommy. When not absorbed in study, he preferred to explore some nearby earthworks which were the remnants of a Roman fort built around AD 80. Tommy, a keen student of Latin and history, was fascinated by these ruins which were spread over eight acres. The fort was constructed on the orders of the Roman general, Gnaeus Julius Agricloa, and marked one of the most northerly points of Roman settlement in Britain.[7]

For the first eight months after Britain's declaration of war, the Western Front was quiet as Adolf Hitler ground Poland into submission. It was during this so-called 'phony war' in the west that many evacuees returned home. Thomas and Mary Reid, however, stood firm, partly because Tommy had settled into Morrison's, but mainly because Thomas believed that the massive British warships he was working on would one day make John Brown's shipyard a Luftwaffe target. It came as no surprise to Thomas that on the evening of 13 March 1941, all hell broke loose when 236 German aircraft dropped almost 300 tons of high explosives and 1650 incendiaries onto Clydebank. Because some of the German pilots mistook the long gentle curves of Great Western Road for the River Clyde, many of the bombers missed the shipyards altogether, unleashing their payloads on Clydebank's residential areas. The following night the Luftwaffe returned. When it was all over the official toll was 528 people killed and 617 seriously wounded, with all but seven of Clydebank's 12 000 houses badly damaged. Unofficial estimates put the death toll at 647 with 1680 injured. A number of those killed had no visible injuries. The bombs' shock waves had been enough to destroy their brains. Some of the bombs had been dropped further east over Glasgow, which also sustained serious damage and numerous casualties. During the second attack Thomas Reid had two close shaves. The first occurred when a cylinder measuring 8 feet by 3 feet and suspended by parachutes landed around midnight outside his local air raid depot in

Dumbarton Road. Exploding sometime later on a delayed action fuse, it demolished the depot and surrounding buildings, killing four people. A slightly smaller device landed in the Reids' backyard at 8 Crathie Drive. But luckily it failed to detonate.[8]

The Luftwaffe returned three more times to bomb Clydebank and Glasgow, on 7 and 16 April and 6 May 1941, wreaking further devastation and loss of life. Having decided that Tommy, June and Bobby should not return home, Thomas and Mary Reid were in no doubt that they had done the right thing. And when Ardoch House closed down a few months later, they arranged alternative accommodation for their children and new schools for June and Bobby, 10 miles to the north in Crieff. For Tommy this meant that he could walk to Morrison's, provided he carried a regulation gas mask whenever he was off campus.[9]

Although Crieff was far removed from the carnage of Clydebank and Glasgow, there were constant reminders of the war, among them an influx of parents with school-age children in search of a safe haven and a good education. Throughout the war Morrison's continued to burst at the seams. Unprecedented enrolment levels tested the patience of the staff, many of whom were temporary teachers brought in to replace full-timers who had gone off to fight. Air raid drills regularly disrupted classes as students raced to take shelter in the school's corridors. Fuel of all types was rationed and during the bitterly cold winters of 1940 and 1941, students shivered through their lessons. A very heavy snowfall on 6 March 1942 forced the school to close on that day. At various times, plans were floated by the military to use the school buildings, in one instance as a 'shadow hospital'. Although this came to nothing, over 400 men of the 157th Brigade did at one stage occupy a large part of the school for sleeping-quarters. Meanwhile cutbacks to the number of teachers and shortages of paper and fuel had led to Scotland's national examinations being organised at a local level. And certificates were issued to meritorious students in lieu of academic prizes, with the money saved going to war charities. Through it all Tommy Reid remained focused on his schoolwork as his academic interests began to crystalise.[10]

What emerged as Tommy progressed through high school was his talent for maths. Back in the 1940s it might have been said that he was 'left brained' – a logical, analytical and objective person. Today's neuroscientists emphasise that the study of maths and science engages two modes of thinking: focused and diffuse. The focused mode, involving the use of rational, sequential and analytical approaches, is essential. It is related to the ability to concentrate, the source of which is found in the human brain's prefrontal cortex located immediately behind the forehead. Almost as important is the diffuse mode because it allows contemplation of the bigger picture, enabling fresh insights into tricky maths or science problems. In Tommy's case, by whatever combination of nature and nurture his mind had come to be what it was and however it worked, one thing was clear, that by his mid-teens he was demonstrating a pronounced flair for mathematics.[11]

If Tommy Reid sometimes came across to his classmates as a swot, it was because of the immense concentration he brought to bear on mathematical problems, putting all else out of his mind as he wrestled with whatever challenging conundrum was in front of him, before coming up with the answer. As a winner of the Fields Medal in mathematics, William Thurston, once put it:

> Mathematics is amazingly compressible: you may struggle a long time, step by step, to work through the same process or idea from several approaches. But once you really understand it and have the mental perspective to see it as a whole, there is often a tremendous mental compression. You can file it away, recall it quickly and completely when you need it, and use it as just one step in some other mental process. The insight that goes with compression is one of the real joys of mathematics.[12]

This process of 'chunking' – the creation of conceptual chunks, which are mental leaps uniting separate bits of information through meaning – helps to develop expertise in science as well as in mathematics. In each case the level of abstract encryption, where a single symbol can represent a number of different operations (such as a multiplication sign also representing repeated additions) adds a level of complexity where the main emphasis must be on focused effort, compared to the humanities where the primary driver is diffuse mode effort.[13]

Being good at chunking in mathematics makes it easier for someone to grasp or create similar chunks in other key subjects. This is especially so in physics, chemistry and engineering, each of which rely heavily on underlying mathematical principles. It was no surprise, therefore, that Tommy Reid's best subjects at Morrison's apart from mathematics were physics and chemistry or that he later became an engineer. It was also no surprise that when Tommy sat for his Senior Leaving Certificate in the summer of 1944, he blitzed his maths and physics exams, correctly solving the following problems amongst others:

> *P* and *Q* are points of contact of tangents *OP* and *OQ* drawn to a circle whose centre is *C*; *PC* and *OQ* are produced to meet in *L*; *QC* and *OP* are produced to meet in *M*. Prove that *LM* is parallel to *PQ*.
>
> An object 2 cm. high is held at a distance of 60 cm. from a convex lens of a focal length of 20 cm. Find the nature, position, and size of the image.[14]

In the early seventeenth century, Galileo famously said that 'the laws of Nature are written in the language of mathematics'. And during his years at Morrison's, Tommy had come to appreciate that this language is both intergenerational and universal: a Pythagorean theorem means the same thing today as it did to the ancient Greeks; and maths is understood by mathematicians the world over, regardless of their different native tongues. When Tommy graduated with distinction from Morrison's in the summer of

1944, he might have had in mind a thought which the renowned mathematician, Edward Frenkel, more recently put this way: 'Mathematics directs the flow of the universe, lurks behind its shapes and curves, holds the reins of everything from tiny atoms to the biggest stars.' At the time of Tommy Reid's graduation, however, the brutal reality was that mathematics was being used to devise ever more destructive weapons, including one which would soon split the atom.[15]

Following Tommy's graduation, he and his siblings returned to Glasgow to live with their parents, who had moved to a more upmarket tenement at 89 Glencoe Street. As Bob Reid recalled it, this place had two bedrooms, a lounge, a kitchenette, hot running water and a built-in enamel bath. But Tommy did not enjoy these luxuries for long. On 28 March 1944, just two weeks after turning seventeen, he had volunteered for the Royal Navy, following which his basic training had been deferred so he could graduate from Morrison's. However, a few days after being awarded his Senior Leaving Certificate, Tommy received a buff-coloured envelope marked 'On His Majesty's Service Official Paid'. Inside was a letter. After categorising him as a 'scholar', it directed him to report for duty with the Royal Navy within three days or face a desertion charge. Being a volunteer, Tommy needed no encouragement and looked forward to the possibility of serving on a warship built by his father's shipyard. As it turned out, Tommy's service took a different direction.[16]

Senior Service Brand

It was Tommy Reid's aunt who first sparked his determination to volunteer. Aged nineteen, Jan Reid had been called up in 1942. Within two years she was on active service in Belgium as a member of the Royal Corps of Signals. Like most of his classmates at Morrison's Academy, Tommy was also keen to do his bit to defeat Hitler. Since the outbreak of war, Morrison's rector, James Donaldson, had been helping Britain's armed forces identify and recruit his brightest students. Like the headmasters of Britain's other elite public schools, Donaldson was constantly on the lookout for those soon-to-be old boys who could rapidly train up to undertake the most challenging technical tasks in the navy, army and air force. The dramatic wartime developments in radio and radar meant that some key combat roles were now beyond the ability of ordinary conscripts. The Royal Navy's 'Y scheme' was designed to cream off the most intellectually gifted volunteers straight from school, to man and maintain the ever-increasingly sophisticated equipment being operated by the fleet.[1]

From the day of his enlistment on 28 March 1944, Tommy was designated an ordinary seaman. But he was also a Y scheme recruit. To get that far he had undertaken a written test and survived an interview with a crusty old retired admiral. He then completed his paperwork which described him as being 5 feet 7 and a half inches tall, with light brown hair, blue eyes, a fair complexion and a scar under his right eye. He was also listed as a volunteer who would remain in the navy 'until the end of the present emergency'. As a Y scheme recruit, he had been immediately released to the Naval Reserve so he could complete his Senior Leaving Certificate at Morrison's. Once he had done so, Tommy received orders to report for duty at HMS *Royal Arthur*. This base was the former Butlin's holiday camp at Skegness on the North Sea coast. Its commanding officer took the salute at each march-past from a 'quarterdeck' which had been constructed partly over a children's playground and partly over a skating rink.[2]

Unlike ordinary recruits, Y scheme volunteers were immediately allocated to specialist units. In Tommy's case, he was mustered directly into the Radio Mechanics Branch as a junior rank telegraphist (wireless mechanician). This branch had been set up in 1942 when the Admiralty decided to equip all its warships with top-secret radar systems. The idea behind the branch was to select the highest quality recruits who demonstrated a talent for technical things and submit them to intense training so that they could service radios, radio detectors and radar arrays on land and at sea. To attract the right people, an elite within an elite, enhanced pay and promotion were part of the package together with a special Y schemers' uniform. While ordinary recruits were issued with traditional bell-bottomed trousers and a striped blue jean collar, Tommy and other Y schemers dressed in a crude version of an officer's uniform including a peaked cap, collar and tie. Although they felt spiffy in their new rigs, one ordinary recruit described them as looking like 'a cross between a taxi driver and a workhouse inmate'. Among his naval peers Tommy Reid was known as Tom. His childhood name fell into disuse.[3]

Under the Y scheme, Tom Reid was on a fast track to 'hostilities only' promotion. If the war continued and all went well with his assessments, Tom would be a leading radio mechanic within twelve months and a petty officer one year after that. It would be some time, however, before he was likely to see any active service because training for the latest radio and radar technology was lengthy and demanding. First, like everyone else who joined the navy, he had to undergo basic training, including foot drill, known in the navy as 'gravel grinding'. For Tom, this meant a move to HMS *Duke* in Worcestershire's Malvern Hills where his only direct connection to life afloat would be rowing a whaler in the camp's swimming pool. Because ordinary recruits were mixed in with them, fast trackers like Tom found this training interminably boring as it could only proceed at the speed of the recruit with the lowest IQ. But what they all learned was vital because on completion, as one officer put it:

> Each could look after himself and his kit; whatever his category, each could swim, and pull and sail a lifeboat; each knew enough about a fighting ship not to be a nuisance at sea. And, above all, they had a sense of belonging, a rock-bottom foundation for living together, in preparation for the time when they would be locked together for months on end in a steel box far from land.[4]

After completing his basic training, Tom settled into the serious business of learning about naval radio and radar. In early 1942 he had read an article about radar in the American magazine, *Popular Science*. It included an artist's impression of a tridimensional radio-locator system: a long metal arm with transmitters at one end, to direct radio waves at an approaching plane, and receivers at the other, to pick up the radio waves as they bounced back. This article had appeared just before America entered the war in December 1941. After that, developments in radar had been bound up in official secrecy. Tom was astonished by the wartime advances in radar, when towards the end of 1944, he finally became privy to a few of them. Some of the Royal Navy's equipment was American, such as the SM-1 height-finder and fighter direction set which had first been fitted to the aircraft carrier HMS *Indomitable* in February 1944. This apparatus could give both plan position out to 80 miles and heights out to 50 miles. When searching, it could sweep continuously, scanning any targeted vertical segment of 12 degrees up to 75 degrees' elevation. It could also detect weather systems.[5]

The Admiralty's enlightened attitude to rapid technological developments was such that after some introductory study of radio and radar, Y schemers were seconded to civilian establishments to study the latest equipment. For Tom and a few others, this meant a twenty-one-week attachment to the General Post Office's wireless transmitter station at Hillmorton in Warwickshire. This facility, which included an array of twelve, 820-foot-high steel masts, had supported the first trans-Atlantic telephone system between London and New York. Now it was Britain's prime centre for communicating with its submerged submarines on patrol and its merchant ships at sea. During his secondment, Tom received lectures from GPO instructors on the theory of wireless communication, while GPO engineers guided him through the practicalities of valves, aerials, transmitters and receivers. Tests in constructing wireless sets, in fault diagnosis and in rectification, were key components of the course which he completed on 10 May 1945, two days after Germany's surrender.[6]

Although hostilities in Europe had ceased, the war against Japan raged on. In anticipation of Adolf Hitler's imminent demise and in recognition of the ongoing threat posed by Japan, the British Pacific Fleet had been formed in Ceylon on 22 November 1944. By April 1945, it was in the thick of the action against Japanese kamikaze pilots. On 6 April, 700 kamikazes based in Formosa struck a US Task Force as it invaded Okinawa, doing great damage. But led by HMS *Indomitable* with its SM-1 height-finding radar, the British fleet was able to destroy the kamikazes' airfields, blunting their attacks. The Japanese were so tenacious in their defence of Okinawa, however, that it was anticipated the war against Japan would last well into 1946.[7]

After completing his secondment with the GPO, Tom Reid participated in a further eighteen weeks' practical work with radio and radar sets. As a wireless telegrapher he studied up on his Morse code because although British warships had radios for ship-to-ship voice messaging, they had limited rage and were vulnerable to enemy interception. In addition to Morse code, Tom had to be across the *Handbook of Wireless Telegraphy*, a tome of almost 1000 pages of highly technical text. Some of the ordinary seamen who enlisted at the same time as Tom had by now been in action on the high seas for months. While Tom must have been frustrated by his seemingly endless study, the Navy insisted on it because all but the largest warships at sea relied on the skills of a single radio mechanic to maintain their communications.[8]

Nevertheless, there were compensations. One of Tom's later training postings, HMS *Scotia* in Ayr, was close enough to Glasgow for him to be able to visit his family. Tom's younger brother, Bob, was awestruck remembering him turning up in his peaked cap proudly smoking a 'Senior Service' brand cigarette. Tom's final advanced training was undertaken at the navy's signals school, HMS *Mercury*, in Hampshire. On 18 August 1945, he completed his transformation from a school-boy into a highly disciplined, professional and skilled technician and one of the most sought-after rates in the Fleet.[9]

Since mid-1945 Tom Reid had been earmarked for active service with the British Pacific Fleet and he fully expected to be in combat for the next twelve months at least. But following the atomic bombing of Hiroshima and Nagasaki in early August, the Japanese quickly surrendered. Nevertheless, fresh naval personnel were still needed in the Far East to replace British combat veterans and to support allied forces now occupying the former Japanese Empire. As a result, Tom made his first voyage overseas, arriving in Australia two weeks before Christmas. Like just about everyone else who has ever arrived in Sydney by ship, Tom was enchanted by its harbour. He didn't get to enjoy it for long as he was transhipped by lorry to HMS *Golden Hind*, a massive mosquito-infested tent city on the outskirts of the south-western Sydney suburb of Liverpool. Baking hot in summer, this holding base was home to over 3000 personnel as they awaited news of their assignments. In April 1946, word came through that Tom would be transferred to Ceylon aboard HMS *Victorious*, a 30 000-ton, state-of-the-art aircraft carrier. Travelling with him on their way to London were a host of British veterans and their Australian war brides. Many years later Tom recalled that it had been a 'happy ship' despite being pounded by a massive storm in Bass Strait, which continued all the way to Fremantle. The voyage was so rough that it hit the headlines with the *West Australian* newspaper describing the worst of it as follows:

> At times the wind speed over her flight deck was measured at over 80 miles an hour. Huge seas caused her to reduce speed to 8 knots...compared to her normal cruising speed of 20 knots. Green seas over her bow swept away guardrails... and the flight deck was out of bounds to all on board.[10]

After arriving in Ceylon, Tom was billeted in yet another tent city holding base, HMS *Gould*, which had been carved out of the jungle north of Colombo. Finally, on 8 July 1946, news of Tom's substantive posting came through: Singapore's naval air station, HMS *Simbang*, which provided land-based support for the British Pacific Fleet's air arm. Within weeks, Tom was promoted to petty officer, assuming responsibility for the maintenance of telegraphic communications equipment on the base. One of the many challenges he faced was maintaining his equipment so far from home, compounded by the American admiral William 'Bull' Halsey's refusal to lend the British any spare parts. It was not unknown among the allies' more junior ranks for bottles of Scotch to be exchanged for American spares.[11]

Tom Reid remained in Singapore for over a year before returning to Britain, where he was formally discharged just before Christmas 1947. Having received a series of glowing assessments from his commanding officer, Captain Jocelyn Salter, Tom left the Royal Navy an accomplished radio technician and skilled junior leader. He was twenty years old. Although he was liable to be recalled for further service, it was only in circumstances of 'extreme emergency'. To aid Tom's re-entry into civilian life, the Navy issued him with a double-breasted three-piece suit, two shirts with matching collar studs, a tie, shoes, raincoat and felt hat. He was also given a buff-coloured ration book. Noting that his total new clothing allowance under the government's rationing scheme was twenty-four points, Tom quickly calculated that the Navy's civilian issue to him which was exempted from the scheme, was more than double that. He would have needed twenty-six points for his suit alone. It was a sign of his life ahead where bacon, butter, sugar, meat, tea, cheese and sweets were all still severely rationed too. As he strode out through Portsmouth's Unicorn Gate for the last time with his kit bag slung over his shoulder, Tom was confronted by the grim reality of row upon row of bombed-out buildings which still remained in ruins six years after the Luftwaffe had last attacked. He noted that people were taking extra precautions for the approaching winter, the one before having been Britain's coldest in 300 years. It was not surprising that as he undertook the long train journey to Glasgow via London, his thoughts turned to Sydney where two years earlier he had briefly enjoyed its magnificent harbour in high summer.[12]

Although 89 Glencoe Street was a large tenement by Glasgow standards, it was a squeeze to accommodate Tom's parents, his eighteen-year-old sister, June, and his fourteen-year-old brother Bobby, as well as Tom himself. But as many of Glasgow's buildings still lay in ruins alternative accommodation was hard to come by. Having become used to living in confined spaces during his time in the navy, Tom was not bothered by the squeeze. Besides everyone was busy. Tom's father was back at Fairy Dyes, June was doing office work at McDonald's biscuit factory, Bobby was at the local high school and Tom's mother, Mary, was keeping house for them all. It was Tom who had time on his hands until he secured a job testing light bulbs for longevity at a specially built Art Deco tower that was part of the Luma Light Bulb Factory. In the evenings Tom would regale his family with stories about

Australia; about Japanese prisoners of war who were now locked in the same Changi prison camp that only a few months earlier had teemed with thousands of emaciated allied prisoners; and about his prowess with all things electrical. After dinner one night, Tom tried to adjust the time on an electric clock. This was the first one his parents had ever owned. But as a result of his fiddling the clock broke. And much to Bobby's amusement his hot-shot elder brother couldn't fix it.[13]

Tom Reid had no intention of testing light bulbs for the rest of his life. He was just biding his time and making a bit of money while waiting to enrol at the University of Glasgow. Founded in 1451, this establishment was the fourth oldest university in the English-speaking world. It had played a key part in the Scottish Enlightenment whose leaders had included one Thomas Reid. In the mid-eighteenth century, Reid had been the university's professor of moral philosophy, succeeding the legendary economic philosopher, Adam Smith, to that chair. Founder of the 'Common Sense' school of philosophy, Reid believed that before people could reason together they had to agree on first principles.[14]

While Professor Reid's unrelated namesake had loads of common sense, his focus was on engineering rather than philosophy. Although the university's school of engineering had only been established in 1840, it was in the words of one of its professors, 'the oldest university school of engineering to be found anywhere'. In 1923, a separate faculty of engineering had been set up, including a chair in electrical engineering. With his high school flair for maths and physics now honed through training on some of the Royal Navy's most advanced equipment, electrical engineering was a natural choice for Tom. In 1948 he duly enrolled in a four-year degree course and began commuting by tram from Glencoe Street to the main university campus about 2 miles away. Located at Gilmorehill above the bucolic, tree-covered banks of the River Kelvin, the university's Gothic revival style buildings clad in blonde sandstone, were landmarks in Glasgow's West End.[15]

Tom Reid's Class of 1952 was a small group of sixteen students. Most of them were older than the average undergraduate, having served in the armed forces and having seen something of the world. They seemed intent on getting ahead rather than on frequenting the nearby bars and taverns, having taken note of the following advice in the engineering faculty's student handbook.

> University study is a serious business which involves much more than merely putting in attendance and submitting required pieces of work. A university student is expected to devote most of his total time and energy to his studies and activities directly relevant to them. This applies even in vacations.[16]

Almost immediately Tom Reid struck up a rapport with his professor, Bernard Hague, who had been appointed to the chair in electrical engineering two years earlier. Aged fifty-five, Hague looked older, partly because of his unruly shock of longish white hair and his old-fashioned double-breasted coats and bow ties. After starting his working life as an apprentice millwright, Hague's natural brilliance earned him an exhibition which had led to a first-class honours degree and ultimately to a doctorate of science. Having learned to work with his hands as an apprentice, Hague was renowned for building ingenious mechanical models to explain complex mathematical propositions to his students. Expanding on this, Hague did his best to dispense with formal proofs being a finite sequence of propositions, each of which is self-evident. In the vector analysis of things like electromagnetic fields, Hague replaced such proofs with an appeal to physical intuition. Although as a talented mathematician and focused thinker Tom was well able to master proofs, Hague encouraged him to engage his diffuse mode of thinking to approach problem solving in a more creative way, a skill which would later prove vital in his work with NASA.[17]

Tom's combination of natural ability and fierce work ethic

soon paid dividends. At the end of each academic year his pre-eminence would be recognised with a string of certificates, awards, bursaries and prizes, culminating in the Howe Prize for 1951–52. Conferred by the university senate, it was the most valuable and prestigious prize in the faculty of engineering and was awarded to the student 'who had attained the highest standard of excellence in the work of the lecture and laboratory classes'. Although Tom had allowed himself an occasional date during university vacations, he had been careful to avoid any sort of serious romantic involvement lest it interfered with his studies. As his brother, Bob, put it, 'marriage was not on Tom's agenda'.[18]

Then one day towards the end of his course, Tom met Betty McKenna.

Lieutenant Sparks

The story of Tom Reid and Betty McKenna began in a biscuit factory which, among other things, made a popular chocolate biscuit called the 'Penguin'. After tasting one an Arnott's food technologist developed something similar for the Australian market; it was called the Tim Tam. Penguins had been invented in 1932 by William McDonald, whose Glasgow biscuit business had been absorbed into the United Biscuit Company in 1948. A few months before this takeover, Tom's sister, June, had signed on as a clerk at the company's head office. Described by her aunt, Jan Reid, as a 'crazy happy person', June was different to Tom in almost every way. As their brother, Bob, put it, 'June was very glamorous and liked to wear nice clothing. She was light hearted and happy-go-lucky. She liked to drink and have a good time.'[1]

Sitting not far away from June in United Biscuits' head office was a twenty-two-year-old comptometer operator, Elizabeth McKenna, who was known to her family and friends as Betty. Comptometers were state-of-the-art machines used to calculate wages, costings, production, purchases and budgets. Apart from standard additions, subtractions and multiplications, they could calculate things such as percentages on discounts and mark-ups, and interest on loans. The keyboard of a comptometer typically consisted of eight or more columns of nine keys each. To obtain her operator's certificate, Betty McKenna had undergone rigorous training and testing which included high levels of manual skill, mental arithmetic and business knowledge.[2]

In any large Glasgow factory like the one which churned out Penguins, there were bound to be Catholics and Protestants working side by side. But the divide was stark. 'You didn't get choices in Glasgow,' Bob Reid recalled. 'You were either orange or you were green.' And you could be judged by what you ate. As one factory worker put it:

> Catholics didn't eat meat on a Friday. For the new guys on the job, Friday was the day. They're eating their sandwich and everybody is looking to see what's inside it...If you were Protestant and you happened to like cheese you made a point of not eating it on a Friday, in case they thought you were Catholic.[3]

Like their father before them, Tom, June and Bobby Reid had regularly attended a Protestant Sunday School when they were young. On his enlistment papers Tom had noted his religion as 'Church of Scotland'. But none of them was deeply religious. By contrast their mother, Mary, was far more devout and regularly attended the Free Church of Scotland, a breakaway evangelical group known as the 'Wee Frees'. While Mary had strong views about Catholics, her husband and children harboured no such biases. Betty McKenna's Catholic faith was therefore no obstacle to her striking up a friendship with June Reid. Betty became a regular visitor to 89 Glencoe Street.[4]

Of June's brothers, it was eighteen-year-old Bobby who first set eyes on Betty. Over sixty-five years later he could clearly recall the moment, saying simply, 'I thought she was gorgeous.' When Tom finally met her, he thought so too. Dark-haired and fair-skinned with a button nose and a petite, shapely figure, Betty had a bubbly personality which was offset by a more serious almost enigmatic side. Like Tom she was an avid reader. But even though they attended dances together at the university's student union and there was a strong attraction between them, Tom remained intent on his studies while Betty continued to date others.[5]

All the while, Tom's parents had been considering their future. Soon after the war Mary Reid's brother, James Boyd, had emigrated to Launceston, Tasmania. In his letters to Mary, James painted a rosy picture of far-off Australia when compared to Glasgow, parts of which were still bombed out ruins. Even more irritating for Glaswegians was the ongoing tyranny of the ration book, which was still necessary to buy petrol, food and clothing. At the beginning of 1952, Tom's father, who was by now almost fifty years old, made the momentous decision in conjunction with Mary, to emigrate to Launceston where he had secured a job as a maintenance engineer with Tasmanian Government Railways. In March that same year, he and nineteen-year-old Bobby set sail for Australia aboard the RMS *Strathmore*. 'As I went to board,' Bob Reid later recalled, 'a customs officer yelled out to me, "Oy, leave me your ration book" and I threw the wretched thing at him with pleasure.' The others remained behind: June because she was at heart a Glasgow girl, Tom because he needed to complete his degree and Mary because she was determined to ensure that Tom completed his degree.[6]

With his graduation as a bachelor of science in electrical engineering in sight, Tom began spending more time with Betty McKenna. Ever since their first meeting there had been a powerful chemistry between them. As they got to know each better, Tom discovered that he had much to learn. Even though the Royal Navy had broadened his horizons he had not suffered like Betty had. Whereas Tom had remained safely evacuated from Glasgow during the Luftwaffe's attacks, a homesick Betty had returned to her family home in Clydebank. And on the night of the Clydebank Blitz her house had received a direct hit. Having been taken to a nearby shelter when the air raid sirens sounded, twelve-year-old Betty had had some protection from the fearful blast. But she never forgot its ear-splitting sound, its massive shock wave or the dust and debris which rained down everywhere around her. With the shelter suddenly plunged into darkness, Betty was momentarily blinded, concussed and disoriented. Her first thought was that she might be entombed alive. When she was finally able to crawl outside there was destruction all around. Thankfully none of her family had been killed or seriously injured.[7]

As a young boy Tom had watched the slow physical decline of his great-grandfather. But this was nothing compared to Betty's experience with her sister, Mary. A year younger than Betty, Mary had been her school's sports' captain and was enjoying her first year at university when she was struck down with tuberculosis. Suffering from high fevers, her weight plummeted as she lost her taste for food. Soon she was wracked by fits of uncontrollable coughing, frequently vomiting up blood. As she wasted away, there was nothing Betty and her family could do except to make her as comfortable as possible. Mary's death, when she was just nineteen, severely affected them all.[8]

In sharing these intimate stories Betty and Tom were drawn ever closer together. While Tom's mother, Mary, thought Betty was a lovely young lass, she remained concerned about how their budding relationship might impact Tom's final months of study. In addition to cooking and cleaning for him at Glencoe Street, Mary kept an eye on his social life but her concerns were unfounded. Tom topped his final year, receiving the following letter from Professor Hague:

> It has been a great pleasure to have had you in the department as a student and I take this opportunity to congratulate you on your distinguished undergraduate career and its highly successful termination. This has given great satisfaction to us all.[9]

Tom also had the respect of his fellow students if their choice of quotes beside his photo in the 1952 Engineering Year Book is any guide: 'Crafty men contemn [scorn] studies, simple men admire them, and wise men use them' – *Francis Bacon*; 'Let every man's hope be in himself' – *Virgil*.[10]

In the context of Betty's relationship with Tom, Mary Reid harboured another concern – Betty's Catholicism. As Bob Reid recalled, 'my mother was not impressed' and this was conveyed back to Betty via June. All the while Tom had been mulling over his future. With his outstanding course marks, an academic career was a real possibility. And he toyed with the idea of starting a PhD. But another option presented itself when a coterie of Australian naval officers came to Glasgow in search of recruits with technical skills. What was on offer to Tom was a five-year short-service commission as an electrical lieutenant; a 'lieutenant sparks' as this rank was colloquially known within the Navy.[11]

One day as Tom was mulling over his options, Betty surprised him with the news that she was pregnant. The birth of their child was expected in December. Although Tom had fallen in love with Betty, the thought of marrying her had not yet crossed his mind. However, the prospect of fatherhood changed everything. Within weeks Tom had proposed and Betty had accepted. Any plans Tom might have had for further study and an academic career were abandoned as he refocused on finding a paying job to support his family. The technical skills Tom had picked up in the Royal Navy and his outstanding university degree gave him job options in Glasgow. He and Betty discussed these, but Tom was tired of the relentless drabness of Britain, its winter cold and the ongoing tyranny of its ration books. The visions of Sydney's breathtaking harbour he had glimpsed six years earlier kept playing his mind. And he knew that that was where the Australian fleet was based. For Tom, there was also something else. From what he'd seen of Australia, he believed that his children would have a better future there than any they could hope for in post-war Britain. Almost 60 years later Tom's elder daughter, Marg said at Tom's wake:

> Dad reminded us constantly of how privileged he was to be bringing up his family here [in Australia] and to remind us

> just how tough other people's lives can be. He taught us not to take these privileges for granted, but to appreciate what we have – and continue to work toward a fulfilled life given that we have this great opportunity.[12]

For Betty, the move to Australia would mean living half a world away from her parents and siblings, in a country she had never visited. But part of her wanted a fresh start too. And she loved Tom, who seemed intent on emigrating Down Under. She agreed to go knowing that the Royal Australian Navy wanted her fiancée to commence duties at HMAS *Cerberus* in Melbourne before the end of the year. This meant that there would only be time for a brief marriage ceremony before Tom's departure and that Betty would have to remain behind in Clydebank with her family until their unborn child was old enough to make the long voyage to Australia with her. Once the decision had been made events moved quickly. On 20 September 1952, Lieutenant Thomas Reid RAN married Elizabeth McKenna in Glasgow's registry office. And six days later, Tom boarded the SS *Orontes* for the long voyage to Australia.[13]

According to Reid family legend, Tom arrived in Melbourne just prior to the running of the 1952 Melbourne Cup. He watched in astonishment as Australia's second largest city came to a standstill to listen intently as *Dalray* charged forward from the tail of the field to beat the 200/1 outsider, *Walkin Sun*, by a nose. According to the organisers, a record crowd of 90 000 saw the running of the race and a further record, of 92 000 pounds, was wagered on it. As hard-drinking punters celebrated their wins or drowned their sorrows in bars across Melbourne, Tom found it hard to believe that the Navy he was now part of was again at war and that the 320 crew members of the destroyer HMAS *Anzac* were at their battle stations off the Korean coast, alongside the 175 crew of the frigate HMAS *Condamine*. Tom's first posting, however, was to be at a naval facility that 'never weighs anchor'.[14]

Located 45 miles south-east of Melbourne on Western Port Bay and known less formally as the Flinders Naval Depot, HMAS

Cerberus, a sprawling campus-like base with well-cut lawns, flowering shrubs and miles of roads named after famous seamen, was the major training base for all naval recruits. Within its grounds, a campus within a campus, was the Royal Australian Naval College where future naval officers undertook a four-year course before being commissioned. There was, in addition, a small cohort of recruits like Tom; they had already been commissioned on probation for five-years' short service but needed to undergo an accelerated course combining elements of both basic and officer training. This group comprised people with professional qualifications in engineering, medicine, dentistry or law. One of Tom's colleagues was a fellow electrical engineering graduate, Fred Lynam, who would eventually reach the rank of rear-admiral and chief of naval technical services.[15]

Although Fred Lynam's first impression of Tom was of a 'terse, snippy Scotsman', they soon became good friends. While the others listened intently during introductory lectures on seamanship and such like, Lynam noticed that Tom, who knew this stuff inside out from his time in the Royal Navy, would be sitting quietly up the back working on a calculus problem he had set himself. 'Tom was very independent minded and was resistant to some military issues,' Lynam recalled. 'I don't think he saw the Navy as a full-time career.' Tom's other preoccupation was Betty. As best as the communications of the day would allow they kept in touch. Then just before Christmas 1952, Tom received the wonderful news that Betty had given birth to a son on 20 December whom they immediately agreed would be called Thomas – Thomas the eighth – in the Reid family tradition.[16]

It wasn't until well into the New Year that Betty felt comfortable making the long sea voyage to Australia, arriving safely in Melbourne with baby Tommy aboard the RMS *Orion*, on 1 April 1953. After an emotional and deeply satisfying reunion Tom showed his wife around HMAS *Cerberus*. From his descriptions in letters, Betty had some idea of what to expect. And she was pleasantly surprised by what she saw: a campus akin to a country town by the sea, with its own railway station, shopping centre and hospital. Something that she probably hadn't fully appreciated and which came as a pleasant surprise, was Tom's status as an officer, outranking most personnel on the base. This brought with it some of the best married quarters as well as the respect accorded to an officer's wife. Although Betty keenly missed her parents and siblings back in Glasgow, her living conditions on the base and her reunion with Tom more than made up for her homesickness. On 8 March 1954, Betty and Tom's daughter, Margaret, was born at the nearby Frankston Hospital. While Betty kept her Catholic faith, it was agreed that Tommy and Marg would be raised as Protestants, a decision that helped to draw Betty and Tom's mother, Mary, closer together. Over the next few months, most of which Tom spent at sea, Betty and her infant children would sometimes journey across Bass Strait to stay with Tom's parents in Launceston.[17]

After completing his course at HMAS *Cerberus* in June 1954, Tom was posted to HMAS *Vengeance*, a ten-year-old light aircraft carrier which served as a fleet training ship. Over the next four months he was put in charge of a team responsible for this warship's heavy electronics, including its main generators and its engine room's electrical machinery. Tom was one of a small number of electrical lieutenants, each responsible for a part of *Vengeance's* overall electrical system. At the end of Tom's posting *Vengeance's* commanding officer, the highly regarded Captain Otto Becher DSC and Bar, noted that Tom 'has conducted himself to my entire satisfaction [and is] a loyal and capable officer who is recommended for a permanent commission'.[18]

Tom's next posting was to HMAS *Leeuwin* in Fremantle as Port Electrical Officer, Western Australian Area. This was a substantive command which lasted for two years. Betty, Tommy and Marg also relocated to Fremantle where the Navy allocated them a roomy California bungalow with a generous private backyard. Because *Leeuwin* was another naval facility which never weighed anchor, Tom and Betty and their children had time for regular visits to Perth's Kings Park, to Cottesloe Beach and to the Swan River, where on Navy open days they were able to putter about in work boats. It was a very happy time for all of them. At the end of his posting in September 1956, Tom's commander rated him 'a most capable officer with initiative and drive'.[19]

Tom Reid's short service commission had one year left to run. His next posting, to the 2000-ton Tribal Class Destroyer HMAS *Warramunga* as its electrical officer, was a quantum leap in responsibility. For the first time in his career Tom would be in charge of all the electronic equipment on one the Navy's front-line fighting ships. Launched in 1942, the '*Munga*' as she was affectionately known, had seen action during both World War II and the Korean War. In late 1954, she had undergone a major modernisation. After this refit her main armament consisted of four upgraded 4.7-inch guns, two 4-inch guns and a brand new 'Squid' triple barrelled depth charge mortar. She was also armed with eight anti-aircraft guns and four torpedo tubes. To the casual observer the *Munga* bristled with a forest of weapons' barrels and communication masts, all of them relying on electronics for their operation.[20]

As the *Munga's* electrical officer, Tom Reid was one of four departmental heads, the other departments being marine engineering, supply and seamen. Tom and his three colleagues answered to the warship's captain, Commander Anthony Synnot. Out of the *Munga's* total crew of 250 men, Tom was in charge of seventy personnel, many of them highly trained technicians ranging in rank from a chief petty officer to ordinary seamen. Tom and his team were responsible for the *Munga's* generators and their emergency alternative; a variety of its equipment including the gyro compass, operation room displays, bridge instrumentation, alarm systems, lighting, time keeping equipment, communications, VHF and HF radio, sonar, ship's log, echo sounder, wind speed and direction equipment, its combat systems, its sophisticated Type 293 long-range radar and navigational radar, its computer elements used to aim its guns and last but not least, its equipment used to screen films to entertain the crew.[21]

Tom's responsibilities included the welfare of his sailors, such as their defence at summary punishment sessions conducted by Commander Synnot. Having started as a seventeen-year-old ordinary seaman in the Royal Navy, Tom understood his men and enjoyed sharing a beer with them during shore leave. But he also had a steely side as his younger brother found out in a Launceston pub while sharing a drink with a fellow football player, a hooker called Steve. When Steve said that he had been in the Navy, Bob Reid replied that he had a brother in the Navy. 'He's a lieutenant sparks,' Bob said, 'Lieutenant Reid.' Colouring up, Steve replied, 'I know that bastard!' And then he produced a pink discipline slip out of his wallet, whereupon Bob confirmed Tom's signature at the bottom. After a few more beers Steve calmed down. 'I probably deserved it,' he said.[22]

There were two highlights to Tom's tour of duty aboard the *Munga*. During late November and early December 1956, the destroyer acted as guard ship for the Melbourne Olympics' yacht racing. She was moored in Port Phillip Bay, forming one end of the starting line for what, in the case of the 5.5 metre yachts, was a course 14 nautical miles long. Aboard her were the race judges who signalled the firing of the starting gun and who supervised the races. Much of the equipment the judges relied on was Tom's responsibility. But while he was kept very busy during each race, he still managed to catch glimpses, especially of the yachts' colourful downwind spinnaker runs. In the end, the Swedes won gold, the British silver and the Australians bronze. For the first time Tom had quietly barracked for Australia and for its 5.5 metre skippered by Jock Sturrock. The other highlight for Tom was the deployment of the *Munga* to the Far East in April 1957 to participate in joint exercises with forces of the South-East Asia Treaty Organisation, a body which had been formed in 1954 to provide a bulwark against communist aggression. At the end of Tom's tour, Commander Synnot, who later became a full admiral, a knight, and chief of Australia's defence force staff, reported, 'Lieutenant Reid has conducted himself to my entire satisfaction. Thorough and painstaking in his work, he could always be relied upon to produce good results. [He] took charge of his department well.'[23]

At that time Tony Synnot was already one of the Navy's most respected officers. And his report assured Tom of a lifetime career in the Navy had he wanted it. But Fred Lynam had assessed his classmate correctly. Tom wasn't interested in operating weapons systems. He wanted to be involved in their development and testing. His focus had turned to a desert town called Woomera where a variety of experimental missiles were being launched and tracked. Having enjoyed the delights of living in Sydney for twelve months, Betty was hesitant about moving to the outback but with Tom fixated on his next career move, she eventually agreed.

Woomera

Looking out across the stony treeless desert, Betty Reid now almost two months' pregnant, must have wondered what on earth lay ahead at this place called Woomera. Sitting beside Tom as he concentrated on negotiating the pot-holed dirt road which stretched through the middle of the surrounding wasteland to the horizon, Betty was thankful that four-year-old Tommy and three-year-old Marg had finally fallen asleep among the family's jumble of possessions on the back seat. It was many hours since they had begun heading north into the desert, from Port Augusta on the South Australian coast. And Woomera was still nowhere in sight.[1]

Although only a couple of weeks had passed, it seemed like an eternity since Tom Reid had left his final naval posting in Sydney, where his family had been able to mess about in boats whenever the Navy organised an open day on the harbour. According to their road map, the pitiless country they were in was surrounded by lakes many times Sydney Harbour's size. On closer inspection, these were encrusted with shimmering layers of salt and the only moisture they contained was captured by the evil black ooze lurking below. In the blast-furnace heat, a person without water would die within a few hours. The only water nearby was firmly encased in a concrete-lined steel pipe with a 10-inch diameter which ran almost 270 miles from the Murray River to Woomera.[2]

It was a guardhouse beside the dirt road that provided the first hint Woomera could not be far off. After producing identification, Tom and his family were waved through and a few miles further on, Woomera itself came into view. The most striking thing about this settlement of 4500 people, 'Woomera Village' the locals liked to call it, were the trees which as if by magic sprouted out of the stony ground. These public plantings were spaced at regular intervals along the sides of Woomera's geometrically patterned streets that had all been bulldozed out of the desert within the last ten years. Before then, the site of the village had been part of the surrounding treeless desert. The only thing keeping Woomera's greenery alive was reticulated sewerage effluent. After driving down Banool Avenue, the village's main thoroughfare, Tom turned onto Booromi Street and pulled up outside what was known as a Riley-Newsum, a prefabricated house with weatherboard walls and a ribbed aluminium roof. For the next couple of years this would be the Reids' home until a more substantial brick house became available nearby. The rental was thirty-five shillings per week, added to which was a small fee for furniture and white goods including a refrigerator and washing-machine. Evaporative air-conditioning and later on, air-conditioners, made the Riley-Newsums habitable during summer when outside temperatures above 48°C (115°F) were not unknown; a crinkly-dry heat which often made it hard to breathe.[3]

Organised along military lines, Woomera was a closed community and no one was allowed into or out of the village without a pass. All residents, including children, had to carry their identification with them and had to produce it on demand. Like a military base, there were separate messes or clubs, the junior staff mess being located at one end of town and the senior staff mess at the other. Where you lived depended on your rank or professional standing. As a former naval lieutenant and now designated as a scientific officer grade three in the Commonwealth public service, Tom automatically became a member of the senior staff mess. And his house was located in an area which was generally considered the better end of town.[4]

Woomera Village owed its origins to Wernher von Braun. As a thirty-two-year-old Nazi and member of Adolf Hitler's Schutzstaffel or SS, von Braun had developed a ballistic missile, the V2 rocket, which had rained death on London during the closing stages of World War II. Almost 46 feet tall, 6 feet in diameter and weighing a total of 14 tons, this monster could travel a distance of almost 200 miles at a speed of 4000 miles per hour to deliver a warhead weighing one ton. In the process of doing so, its trajectory reached an altitude of over 50 miles, making it the first artificial object to cross into space. Once it had blasted off from its launcher, there was no way it could be shot down, deflected or otherwise destroyed. Realising that combined with an atom bomb, such a missile would change the face of warfare forever, and unwilling to completely rely on the United States for protection, the British government resolved to develop its own ballistic missile and atomic deterrent. If a missile could not be shot down then the nation threatening to use it had to know that the country it was targeting could respond with equal ferocity. In the age of atomic weapons this became known as the doctrine of Mutually Assured Destruction (MAD). However, in terms of rocketry, Britain was starting this arms race at a disadvantage, von Braun and many of his team having surrendered to the Americans while the Soviet Union captured most of the others.[5]

Within the British Isles, there was only one place, Aberporth in Wales, where missiles could be fired safely and only then, over short distances. The problem was that they would land in the Irish Sea where recovery was impossible. At that time, scientists needed to be able to recover the remains of a rocket to figure out how it had performed. What was needed was a sparsely inhabited continental range somewhere within the British Commonwealth. After the early elimination of Canada because of recovery problems in the Arctic, the British and Australian governments agreed on a range head near what later became Woomera, with a centre line stretching over 1000 miles to Eighty Mile Beach on the north-west Australian coast. The other key limb of the agreement reached by the Atlee and Chifley governments in 1947 and known as the Anglo-Australian Joint Project, was the use of a former munitions complex at Salisbury outside Adelaide as a centre for rocket-weapons research and development. In 1955, the various entities which occupied this site, namely the Chemical and Physical Research Laboratories, the High-Speed Aerodynamics Laboratory and the Long-Range Weapons Establishment, were combined to form the Weapons Research Establishment or WRE for short. And what was developed by WRE was tested at Woomera.[6]

During Woomera's first ten years, nine separate ranges operated nearby, among them Range A for bomb ballistics and Range G for short missiles. By the time Tom Reid arrived in 1957 only Range E remained. It was the best equipped land range outside the United States and the Soviet Union. American observers agreed that 'there wasn't a range in the western world better equipped or better managed or that conducted a major program on fewer staff'. Located approximately 40 miles west of Woomera, the Range E complex was heavily guarded. Most staff arrived from Woomera on large articulated buses comprising a prime mover and passenger trailer with seating for sixty. At the guard house which straddled the only entrance to the heavily fenced complex, each passenger had to produce their photo-identification. Anyone who couldn't do so was denied entry. The largest structure beyond the guard house was the instrumentation building which contained the key control, communication and recording equipment for trials. Nearby were the launching aprons, some of them with fixed launchers, and to their side were heavily reinforced block houses equipped with monitoring devices. Out the back of the instrumentation building were the test shops where missiles were prepared after arriving from WRE Salisbury, together with a measurements shop, an oxidant filling post and a magazine for storing explosives. Stretching off beyond the fenced area down each side of the range were theodolites, telescopic high-speed tracking cameras, Doppler systems for determining speed, tracking and surveillance radar, and aerials and receivers for telemetry. In the wake of the Petrov affair which had disclosed a snapshot of the KGB's reach into Australia just three years earlier, security was tight. At times it went over the top, as with a plan for a three-seat latrine comprising a wooden shelter over a pit, which was classified 'Secret'. When someone protested that it was only an earthen drop toilet, the engineer who had stamped the plan 'Secret' explained: 'But it's in the Tech Area and everything going in there has to be classified.'[7]

The business of Range E was to test missiles. But the tangled, burnt-out wreckage of a spent projectile fired from this range was of limited use in determining how it had performed. A pile of scrap metal could not tell engineers much about what had worked and what had failed. It was crucially important to be able to track an experimental missile in flight and to figure out how its systems had performed while it was airborne. To work out the velocity and trajectory of a missile, Range E had Doppler radar. This equipment was based on the Doppler effect: if you stand by the side of a road and a fire engine races towards you, the pitch of its siren gets higher as it approaches you and lower after it has passed you. But for a firefighter on board, the pitch remains constant. Utilising this effect, Range E's reflection Doppler system deployed a steady signal transmitted from a ground station to a missile in flight. This signal was then reflected back to the ground from the missile, with the different frequencies between the two signals providing the necessary data to determine velocity and trajectory.[8]

Equally important for figuring out a missile's performance was telemetry, literally the remote measurement of things. Applied to a missile in flight, telemetry involved sensors in the missile measuring properties like heat, volume and pressure and changing them into electrical quantities. These quantities could then be transmitted over considerable distances to a Range E receiver which converted them back into measurements. A simple example of line telemetry is a fuel gauge in a car. A float in the fuel tank at the back of the car is connected to a resistance at the top of the tank. As the fuel level varies, the resistance changes too, permitting more or less current to flow along a wire which connects the tank to the car's fuel gauge, thereby changing the readout on the dashboard.[9]

Tom Reid's official designation as 'officer-in-charge of telemetry and Doppler ground instrumentation at WRE Woomera' placed him at the centre of Range E's experimental work. On the face of it this was a meteoric rise for someone whose last job had been that of an electrical lieutenant aboard HMAS *Warramunga*, a fifteen-year-old warship with dated weapons systems which was due to be decommissioned by the end of 1959. What made Tom's promotion even more astonishing was the fact that the systems now under his control at Range E were due for massive change. One of the problems with reflector Doppler was that it tended to measure everything in the sky indiscriminately; not just the vital missile but also its discarded boosters as they tumbled back to Earth. To address this, a new transponder Doppler system was about to be introduced, where a small transceiver in the missile would pick up the Doppler signal from the ground station, thereby avoiding any confusion with its boosters. Equally in telemetry, as radio valves were replaced by transistors and transistors were in turn replaced by tunnel diodes which were a type of semiconductor, it would soon be possible to transmit 400 different measurements from 400 different sources. So what was it about Tom that had led to him being given the job of OIC Doppler and Telemetry at Woomera?[10]

Among those involved in Tom Reid's selection were the soon to be director of the WRE, Bill Boswell, who among other things had helped to install the first radar in Australian naval ships. A charismatic bear of a man with a high public profile, Boswell was known to the media as 'Mr Rocket Range'. Another selector was Mervey Stewart Kirkpatrick, a protégé of Boswell's who preferred to be known as MS Kirkpatrick or just 'Kirk' because his Christian name, which rhymed with 'nervy', was an unfortunate coincidence for a rocket scientist. After developing radar receivers for ground stations in Britain, Kirkpatrick had become a radar and radio Doppler researcher at the WRE in South Australia. Soon afterwards he was promoted to the post of principal officer in WRE's Electronic Instrument Group, where he developed a reputation as an inspiring no-nonsense leader and a shrewd administrator. In considering Tom, Boswell and Kirkpatrick were mindful that if selected, he would need to manage a diverse team of self-taught or service-trained practical people in a variety of specialties including radio, electronics and engineering, as well as certified technicians and experienced university graduates. He and his team would also be required to adapt to the latest fast-moving developments in technology. In reality it was the totality of Tom's resume which made him stand out: his experience as a service-trained radio technician in the Royal Navy which had seen him promoted from a seventeen-year-old ordinary seaman to a petty officer radio mechanic by the age of twenty; his brilliant academic record at Glasgow University where he topped his year with a first class honours degree in electrical engineering; and his service aboard HMAS *Warramunga*, where as a lieutenant, he had led that ship's electrical department of seventy sailors and received outstanding assessments from his captain. In short, Tom combined a brilliant and innovative mind with a capacity to lead men.[11]

Among those on Reid's team was a twenty-two-year-old technician, Bill Miller, who many years later would rise to become the director of technical services at Australia's leading security and intelligence agency, ASIO. After joining the Queensland police force as a seventeen-year-old cadet, Miller had elected to specialise in police radio communications. Based in Cairns, he spent more time than he cared to in pubs, 'as cops do up there'. In 1958 he answered an advertisement for a job at Woomera where the most exciting developments in communications were taking place. Posted to Range E, Miller was assigned to telemetry on Black Knight rockets. Almost sixty years later, he recalled the Tom Reid he knew there.

> Tom was an exceptionally brilliant man, and exceptional mathematician who was always right. He was an exceedingly well educated brainy guy. At times, he was unpopular with some of his colleagues because he was too smart. Even so he was good with people and didn't shout them down or 'put them in prison'. He knew how to motivate people, especially the many ex-navy technicians who reported to him. He didn't knife people in the back because he didn't know how to. And none of his team ever said: 'that bastard Tom Reid'.

While at Woomera, Bill Miller had continued to take courses in electronic communications. One day he came to Tom with a calculation problem that had stumped him during a test set by the Radio Institute. 'This is what the buggers gave me for an exam,' Miller said. 'Leave it with me,' was the reply. Within six minutes, Reid had returned with the correct answer.[12]

Tom Reid's arrival at Range E coincided with plans to fire the first Black Knight, a two-stage missile whose first stage rocket would propel its second stage dummy warhead to an altitude of 500 miles, before the warhead re-entered the Earth's atmosphere at a speed of up to 12 000 miles per hour. The Black Night was a test vehicle for the much larger and more expensive Blue Streak medium-range ballistic missile. Designed to deliver a nuclear warhead onto a target up to 2000 nautical miles away, Moscow and most other major Soviet cities would be well within a Blue Streak's reach. It was pursuant to an Anglo-American agreement struck back in August 1954, that the British had focused on Blue Streak while the United States had proceeded to develop an inter-continental ballistic missile with a range of over 5000 nautical miles. Such rockets were a quantum leap beyond von Braun's V2s in terms of range and altitude. And they presented special challenges as they plunged back into the Earth's atmosphere at unprecedented speeds. The most significant challenge was how a warhead could be protected from the extreme frictional heat, buffeting and shock of re-entry, to reach its target. The same size as a railway carriage, a Blue Streak was three times larger and ten times heavier than a Black Knight. But the re-entry problems posed by both missiles were essentially the same. The much cheaper Black Knight was therefore used at Woomera to save money with the early testing.[13]

Initially scheduled for July 1957, the first Black Knight launch had been delayed when part of the launching gantry was lost overboard while being unloaded from a ship in Port Adelaide. Beyond that, though, the sheer complexity of what was being attempted caused further delays across every facet of Range E's preparations including telemetry. Tom Reid's team had to prepare for the unprecedented flood of information which the Black Knight's various telemetry transmitters would be sending back to Range E, each one on a different frequency. Depending on the precise goals of a particular trial, the telemetry might vary in some respects. But it always included information about the missile's major systems such as propulsion, guidance and control, together with measurements covering acceleration, temperature and vibration. All such data was to be transmitted to the Range E instrumentation building where Tom's team would run equipment to receive and decode the missile's incoming signals which would then be permanently recorded on paper charts and magnetic tape for later analysis back at WRE Salisbury.[14]

It wasn't until just after 8 pm on 7 September 1958 that observers noted a grey cloud of smoke mushrooming from under the first Black Knight missile, codenamed BK01. Nothing else seemed to be happening. It took ten seconds for the missile's ghostly grey nose cone to rise above the smoke. Slowly at first and then with rapidly increasing speed, it roared into the night sky, the four tongues of flame from its engine chambers making a spectacular display for the children of Woomera, most of whom had been allowed to stay up late. While the launch proved to be flawless, the missile's engine shut off prematurely at a height of 40 miles, possibly due to an explosion. It was the telemetry collected by Tom Reid's team which helped to determine what had happened. According to their data, the signals from the transmitter in the rocket's body had been cut off abruptly while the transmitter in the nose cone had sent signals for another six minutes. From this it was deduced that an electrical fault had activated the in-flight-destruct system. Despite this fault, BK01 was publicly proclaimed a success because its purpose had been to test Black Knight's launching systems and not the second stage dummy warhead. More complex tests involving dummy warheads took place throughout 1959 and into 1960.[15]

Up until late 1959, British built AA No. 3 Mk 7 radars had been used to track missiles fired from Range E. These tracking pulse type radars had first been used to aim anti-aircraft guns during the latter stages of World War II. Enemy aircraft were tracked by a narrow aerial inside a dish-reflector mounted on top of a small air-conditioned, four-wheeled trailer. This contraption was about the size of a domestic caravan and contained the radar's key working parts. Once locked onto a target, a Mk 7 could 'feed' that target's coordinates to nearby guns. But as Black Knight tests during the late 1950s became increasingly complex, this mid-1940s technology struggled to track missile trajectories. Keen to assist with the development of Britain's medium range ballistic missile, the Americans gifted Woomera with two FPS-16 radars which were installed in early 1960. Housed in large, specially designed two-storey brick and concrete buildings complete with roof-mounted, fast-moving, twelve-foot diameter dishes, these radars dwarfed the Mk 7's trailers and were the largest instruments ever to be installed at Range E: one of them at Red Lake, 20 miles north-east of the Instrumentation Building, and the other 105 miles west of it, at Mirikata. Like the Mk 7s, these new FPS-16s were tracking pulse type radars. But they were much more precise and versatile. On 31 January 1958, an FPS-16 had been used to guide the launch of *Explorer 1*, the first American Space Satellite, from Cape Canaveral in Florida. It was part of the deal between the British, American and Australian governments that these new radars could be used for space tracking if the need arose.[16]

With the increasing complexity of each Black Knight test being matched by more demanding radar tracking and telemetry, Tom Reid's team faced new challenges with each launch. But there was one consolation; they carried out their work in Range E's instrumentation building, an austere two-story complex, its roof bristling with the latest aerials and antennae. Variously nicknamed 'The Citadel', 'The Taj Mahal' and 'The Tomb', it was stuffed full of the latest electronic equipment worth millions which had to be operated in an air-conditioned and dust-free environment. While just about everyone else in and around Woomera had to put up with searing temperatures, it was not unknown for members of the telemetry team to wear jackets inside The Citadel to ward off its chilly temperature.[17]

For Betty Reid and her children, relief from the heat came primarily in the form of evaporative air-conditioning, dips in Woomera's Olympic-sized swimming pool, and for Nicholas's birth on 17 May 1958, in a purpose-built maternity unit in the village's modern, air-conditioned hospital. Like other children their age, Tommy and Marg were allowed the run of the village by day, if they were not attending the Woomera Area School. Betty, known affectionately around the village as Liz, preferred early morning walks before the mercury rose, during which she would chat to her friends. Many of them were young mothers just like her. Bill Miller remembers her as a happy and engaging community-minded person. Pictures of ordinary life in Woomera are rare because photography was restricted for security reasons. When not in use, family cameras had to be left at the village store. Per capita, Woomera had the highest birth rate in the nation. As the author, Ivan Southall, put it after a visit to Woomera in 1961: 'Woomera's fecundity is almost an embarrassment. Its teeming children are tanned and healthy and bubbling over with life. Its school is superbly equipped...In many ways it is a family town.' If the heat provided one consolation for Betty, it was that by the time she had put the last of her washing on the line, the first item she had pegged would be bone dry. When Tom wasn't up at Range E, he would often put on his old navy whites, shorn of their badges of rank. Dressed in a white shirt and shortish white pants, he'd spend as much time as he could, playing with his children, his only concession to informality being a nifty pair of thong-like sandals. While Tom's colleagues ascribed his rig to his intense pride in his naval career, his family thought it was a certain Scottish thriftiness that drove him to continue wearing his old uniform. Whatever the case, his sunglasses could not hide his generally serious facial expression, a sign that the ever-increasing complexity of his work on Range E was never far from his mind. Despite his job's relentless pressure, Tom Reid had demonstrated great competence as OIC of the Range's Doppler and telemetry team. Bill Boswell and MS Kirkpatrick were well pleased with the choice they had made. As Cold War competition between the United States and the Soviet Union began spilling over into a space race, Tom Reid was well placed to play a key role but not until the Soviets had utterly humiliated the Americans.[18]

Sputnik and *Kaputnik*

After surrendering to the American army at the end of World War II, Wernher von Braun had spearheaded the development of large rockets for his captors, based on his German V2. But despite his best efforts, it was the Soviet Union that launched the world's first inter-continental ballistic missile, a Semyorka R-7, on 26 August 1957. Weighing 280 metric tons, this massive two-stage rocket was capable of delivering a thermonuclear warhead to a target up to 5500 miles away. 'We captured the wrong Germans' was one American general's frustrated response. It wasn't until 17 December that the first successful launch of an American ICBM, an Atlas SM – 65, took place. While these monsters were designed to deliver warheads to obliterate cities on the other side of the planet, it was clear that they could also be used to place satellites in Earth orbit.[1]

For the Soviet Union, the space race had always been about boosting its international prestige, with a focus on the military advantages to be gained from its rockets and satellites. Just weeks after successfully invading Hungary on 4 November 1956, the Soviet premier, Nikita Khrushchev, had boasted to a group of Western ambassadors: 'We will bury you!' To follow through on this threat, Khrushchev had an advantage. As a dictator he was able to centralise his country's scientific and engineering effort, cloaking it in a veil of complete secrecy. The Americans, meanwhile, had become complacent about the technological lead they had enjoyed since 1945, dispersing their rocket development between the three services. The Navy had its Vanguard; the Army its Redstone; and the Air Force its Atlas. In a further complication, President Eisenhower had announced in 1955 that America's satellite program would be civilian rather than military in nature.[2]

As important as rocketry was for America's satellite program, the development of a ground-based space tracking network was no less so. But there were many challenges which Jack Mengel, one of the network's founding fathers, explained:

> Let a jet plane pass overhead at 60 000 feet at the speed of sound, let the pilot eject a golf ball, and now let the plane vanish. The apparent size and speed of this golf ball will approximate closely the size and speed of a satellite 3 feet in diameter at a height of 3000 miles. The acquisition problem is to locate the object under these conditions, and the tracking problem is to measure its angular position and angular rate…[3]

To do so the Americans developed a network of Mini-track stations. While most of them formed a 'picket line' stretching north/south through the Americas, a couple of others further afield were built to help determine a satellite's initial orbit. These stations each operated like a pair of human ears, using stereo-phonics to pinpoint the location of a given signal source. They worked on a frequency of 108 Megahertz. With the Australian government's agreement, the Americans located a Mini-track station about 7 miles north of Woomera Village. Work commenced in early October 1957 with completion anticipated by March 1958. Coincidentally, Tom Reid and his family had arrived in Woomera just as this station's foundations were being laid. But Reid was not involved; for the time being his duties related to the testing of British rockets on Range E.[4]

All the while, the Soviets had been feverishly re-engineering one of their Semyorka ICBMs to carry a satellite into Earth orbit.

Amid the utmost secrecy, it blasted off from a launcher in what is now Kazakhstan on 4 October 1957. On board was a beach ball-sized polished metal sphere weighing 100 pounds, attached to which were four radio antennae, giving it the overall appearance of a giant insect. As soon as the Soviet newsagency TASS was sure that this metal sphere was in orbit, an announcement was made that its signals could be heard by radio receivers set at forty megahertz. In this way, ham radio operators the world over were among the first to pick up the steady beeping signals from *Sputnik*, the first artificial satellite to orbit the Earth.[5]

Although amateurs could tune into *Sputnik's* signal and people under its flight path could see it streaking across the night sky, the Americans' Mini-track picket line, which was set at 108 megahertz, could not pick it up. As American network engineers scrambled to airlift forty megahertz systems to their key stations, one of them was heard to say, 'we were caught with our antennas down'. While President Eisenhower tried to down play *Sputnik's* significance, derisively calling it 'one small ball in the air', this Soviet first was a shock to Americans' national pride and heightened their sense of vulnerability during what was a period of intense Cold War confrontation. Although a West Point graduate and World War II hero, Eisenhower had never personally fired a shot in anger. As supreme allied commander on the Western Front, his skill had been to make his feuding front-line generals, George Patton and Bernard Montgomery, work together. Many Americans now thought that as president, Eisenhower was spending far too much time playing golf. In stark contrast, Nikita Khrushchev, the son of Russian peasants, had participated in and survived Josef Stalin's purges during the 1930s. As a political commissar, he had worked on the front line helping to keep up Russian morale during the long and bloody Nazi siege of Stalingrad in 1942. While Eisenhower was the epitome of an elegant American gentleman, Khrushchev was a loud and ugly looking little scrapper. But he was also a political force of nature. So too was the lanky majority leader of the United States' Senate, Lyndon Johnson. On the night of *Sputnik's* launch,

Johnson had been hosting a barbecue at his Texas ranch. 'Now, somehow, in some new way, the sky seemed almost alien,' Johnson later recalled. 'I remember the profound shock of realising that it might be possible for another nation to achieve technological superiority over this great country of ours.'[6]

Before the Americans were able to get their own satellite into orbit, the Soviets struck again, this time on 3 November 1957 when they launched *Sputnik 2*. On board was a stray dog from Moscow's streets called Laika or 'Little Barker' who survived in orbit until she was put down by an automatically triggered injection to spare her being burned alive on re-entry. Although the Americans derisively dubbed Laika's spacecraft *Muttnik*, they were panicked into rushing the launch of their first satellite aboard a navy Vanguard rocket just three days later, an event which was televised live. After rising just 4 feet off its launch pad, this rocket exploded in a fireball. At the same time its satellite popped off. After landing in some nearby bushes, it began transmitting its signal. America's tabloids immediately dubbed this grapefruit-sized object *Kaputnik*.[7]

All the while, working for the American army's Ballistic Missile Agency, Wernher von Braun had been finessing his Redstone rocket. But it had not been chosen for the Americans' first satellite launch because of public relations concerns about von Braun's Nazi past. After the humiliation of *Kaputnik*, such sensitivities were cast aside. And on 31 January 1958, America's first satellite, *Explorer 1*, was blasted into orbit atop a modified Redstone rocket. Although President Eisenhower was still intent on a civilian-run satellite program, he now realised that much better policy coordination was needed. With the enthusiastic arm-twisting support of Senator Lyndon Johnson, the American Congress passed legislation on 16 July 1958 which established the National Aeronautics and Space Administration as a civilian run operation. And on 17 December 1958, the fifty-fifth anniversary of the Wright brothers' first powered flight, NASA's administrator, Keith Glennan, announced Project Mercury, whereby the

United States would place a man in orbit and return him safely to Earth, preferably before the Soviets did.[8]

Although Keith Glennan only headed up NASA until the end of the Eisenhower presidency on 20 January 1961, he was a first-class administrator who took over the army's Ballistic Missile Agency, renamed it the Marshall Space Flight Centre and put Wernher von Braun in charge of both it and all rocket and spacecraft research. Glennan also incorporated within NASA, the Jet Propulsion Laboratory (JPL) in Pasadena, California, and the Goddard Space Flight Centre in Greenbelt, Maryland. At different times these organisations would come to play crucial roles in Tom Reid's long and distinguished career with NASA; with JPL running the deep space network to maintain communications with far flung robotic spacecraft, and Goddard supervising the spaceflight tracking and data network to support orbital flights. But just on two weeks after Glennan's announcement of Project Mercury, the Soviets launched *Luna 1*. This was the first man-made object to break free of Earth's gravity, after which it flew past the Moon at a distance of 3700 miles. The Americans again scrambled to catch up. Two months later, NASA's *Pioneer 4* passed by the Moon, but at a distance of 37 000 miles. The Soviets then increased their lead. On 12 September 1959, they landed *Luna 2* on the Moon's surface. And within a month after that, *Luna 3* took photos of the Moon's far side. While the American Congress became increasingly agitated by the Soviets' winning streak, NASA's hard heads focused on manned missions that would give their astronauts the best chance of surviving.[9]

Even before Keith Glennan's announcement of Project Mercury, NASA had begun work on its manned space flight network. While a Mini-track station could follow the path of an unmanned satellite, it took several orbits to get an accurate fix. When that happened, all such a station could do was passively receive the satellite's signal. This was not enough for an orbiting astronaut to maximise his chances of survival. He needed real-time conversations with his controllers back on Earth. And they needed telemetry relating to his heartbeat, respiration and temperature, as well as to his spacecraft's systems, together with an ability to issue remote commands to the spacecraft itself. But these communications would be of little use unless they could be carried out in real time; they had to be both instantaneous and continuous. With an orbiting astronaut flying 100 miles above the Earth's surface at 17 000 miles per hour, any one tracking station would only have the astronaut 'in view' and contactable for about seven minutes at most. A worldwide tracking network would therefore be necessary to get anything like continuous coverage, and each station would need the very best radar to immediately find and lock onto the astronaut's spacecraft as soon as it hurtled over the horizon.[10]

The initial assessment was that twenty stations would be needed. But this was soon whittled back to eighteen. Their locations were determined by the standard orbital flight path of manned missions which on a Mercator's projection map, stretched roughly in a straight line from the West Coast of the United States to the East Coast of Africa, before arcing down to the Southern Indian Ocean, across Southern Australia, and then back up to the US West Coast. Among the sites chosen outside North America were: Bermuda; the Canary Islands; *Rose Knot Victor*, a ship in the Atlantic Ocean; Kano and Zanzibar in Africa; *Coastal Sentry Quebec*, a ship in the Indian Ocean; Muchea and Woomera in Australia; Canton Island, now part of the Republic of Kiribati, in the mid-Pacific Ocean; and Hawaii. Although the American space medicine community had recommended uninterrupted voice communication, this was not possible. And the more notable gaps, depending on the precise orbital path, occurred between Woomera, Canton Island and Hawaii.[11]

Located in sandy scrubby country approximately 30 miles north-north-east of Perth, Muchea was purpose built by NASA for Project Mercury. It was chosen as the only command facility in the Southern Hemisphere because it was 180 degrees distant in longitude from Project Mercury's Cape Canaveral launch site. Australia's other Mercury tracking facility, at Woomera, was not co-located with the Mini-track station there or alongside the Range E instrumentation building. Rather it was built 20 miles to the north-east, near Red Lake's FPS-16 C-Band radar which Tom Reid and his team had been using to track British missiles. With a dish that was 12 feet in diameter, this high-precision instrument could track to an accuracy of 23 feet over a distance of 500 nautical miles, more than enough to establish the elevation and range, and therefore the precise location of an orbiting Mercury spacecraft.[12]

Outside the United States, Australia had more say in running its NASA tracking stations than any other host country. Whereas local staff at other foreign stations were led by an onsite NASA-appointed station director who was invariably an American, the Australian government appointed its own on the recommendation of Salisbury's WRE. In practice this meant that the decision on who would run Project Mercury's Woomera tracking station would be made by Bill Boswell and MS Kirkpatrick. And for them there was only one choice, Tom Reid, who had already proven himself using the FPS-16 radar in conjunction with the telemetry equipment in Range E's Instrumentation Building, to track British missiles. With NASA's first manned sub-orbital flight scheduled for no later than 1 June 1960 and its first orbital flight by 1 January 1961, Tom was soon immersed in training. Although he had mastered the complexities of tracking missiles and warheads, the challenges of communicating with an astronaut in orbit were of an altogether higher order.[13]

To get their senior tracking staff up to speed, NASA established a demonstration site on Wallops Island, Virginia, near a rocket launching facility. Linked to a nearby FPS-16 radar, a specially built tracking station was able to simulate Mercury flight conditions with the aid of an aircraft fitted out to mimic an orbiting spacecraft. Tom Reid began regularly commuting halfway across the world for training sessions, at a time when, for most Australians, flying overseas was a rarity. Having taken off from Woomera's airfield on a regular TAA service, Tom's first stop was Adelaide after which he flew to Sydney. From there he would get a seat on Qantas's brand-new Boeing 707, *The City of Canberra*, for the long flight via Fiji, Honolulu and San Francisco to New York. Then it was a short commuter flight to Washington DC and a further bunny hop to Wallops Island. This trip took two days. Despite his jet lag Tom impressed the many senior NASA operatives he now mixed with. Even in the company of his fellow station directors who had also flown in from elsewhere around the world, Tom stood out. Fifty-six years later, George Harris Jr, who in 1960 was the leader of NASA's aircraft simulation group, could recall the impression Tom Reid made on him.

> Me being a Brit and Tom being a Scot was cause enough to sink a few [Scotches] during the evening hours... Tom was always a can-do guy, with a great sense of humour. He was a good leader and very cool under pressure. I count Tom among the people I always felt it an honour to serve with.[14]

For Betty Reid, Tom's long absences were a challenge. Although seven-year-old Tommy and six-year-old Marg were now attending the local school and pretty much able to fend for themselves, two-year-old Nicholas was still a handful while a fourth child was due in just a few months. But this was the norm in Woomera. Expectant mothers with boisterous broods of children and no relatives living nearby, did the best they could while their husbands travelled backwards and forwards to America or Britain on rocket-related business. With no crime to speak of, unaccompanied children were able to roam the village at will and thought nothing of wandering in and out of their friends' houses at all hours of the day and night. It was as if the whole village was home to one giant family. Betty not only got by but made close friendships with other young mothers who faced the same challenges. Like children everywhere, the Reid kids did sometimes get into mischief, whereupon Betty would warn them that they'd be dealt with by their father when he returned. Although Tom had a stern side, he could never bring himself to admonish his children as they were lined up for a sharp talking to after his long trips away. Instead, everyone would dissolve in warm hugs and gales of laughter as Tom parcelled out bundles of the newest American toys, and for Betty, the latest in New York fashion.[15]

Meanwhile out at Red Lake, approximately 1600 feet away from the FPS-16 Radar, a telemetry building was being constructed as set out in the site handbook of NASA's principal contractor, the Western Electric Company. This building resembled a massive oblong box-like farm shed designed to house tractors and combine harvesters. And in its harsh desert setting, it looked totally out of place. Its interior was even more so. Fully air-conditioned and lavishly appointed, it was designed to house all sorts of the latest gear including a relay rack, sequential selector, loop switchboard, acquisition data console, acquisition aid receiver, various controller and monitoring consoles, event and pen recorders, voice receivers, a calibrator and a tape recorder. Outside, two quad helix telemetry and control antennas, one 25 feet high and the other 35 feet high were installed. These were each square rather than dish shaped. And from each, four antenna poles protruded some distance. They were capable of seeing further over the horizon than the FPS-16 Radar but with less accuracy. Ken Anderson, who worked at Red Lake on later Mercury missions, likened them to different types of torches in a darkened factory. The quad helix antennas could cast the equivalent of a wide but weak beam of light which would provide the first illumination. Taking advantage of this, the FPS-16 could then cast a much more powerful and precise beam of light to illuminate a particular object. When it came to tracking, the quad helix antennas picked up a space craft's signals first, sufficient to establish voice and telemetry links but not sufficient to establish the spacecraft's precise location. That was the job of the FPS-16 which was connected or 'slaved' to the antennas to take advantage of the signal they had already acquired. To be able to first pick up that signal, the antennas had to be properly calibrated. So next to them, a much taller collimation tower was built; a tower that could imitate an orbiting space craft, allowing the antennas to be fine-tuned. As Bill Miller recalled:

> The set up at Red Lake was: a small team of Yanks came into the empty building and began installing racks, running cables and generally setting up the skeleton. We got the antennas working and other outside things. The racks were filled with commercially available equipment: tape recorders from Ampex, receivers from Nems-Clarke, power supplies etc. They did a great job which I watched with envy. Tom Reid's team calibrated and got all going. I was responsible for the Ampex recorders and their operation.[16]

Although Tom chose only the best of his Range E team for Project Mercury, his selection methods could be unorthodox. One day as Tom and Betty were loading up their shopping trolley at the Woomera store, they were approached by Bill Miller who asked Tom if he could be given responsibility for the Ampex tape recorders. These registered the performance of the spacecraft's systems and the astronaut's medical condition on magnetic tapes for immediate playback while also providing a permanent record. Taking a moment to think about it, Tom said 'yes' just as the three of them reached the check-out counter. Miller turned out to be a very popular and competent choice. The Americans nicknamed him 'Rhode Island Red' on account of his ginger hair.[17]

Apart from communicating with the spacecraft, the Red Lake telemetry building also had to be capable of contacting other tracking stations as well as the Mercury Control Centre at Cape Canaveral. This it did through a complex system of submarine cables, land lines and radio links. Woomera was joined via Adelaide to Sydney on a landline. Then from Sydney to the Goddard Space Flight Centre in Maryland there were two alternatives: a submarine cable to Vancouver and thence to Goddard via land line; or a radio link to Hawaii, then a submarine cable to San Francisco and finally to Goddard via another land line.

From Goddard, there were landlines to Cape Canaveral and the North American stations. It was also linked by radio to its more exotically located stations such as Bermuda, Grand Canary and Zanzibar. As between Woomera and Muchea, the link was via a land line which passed through Adelaide and Perth. For the first time in history, a real-time, worldwide communication network with reliable high-capacity data links came into being with the completion of the last station in Kano, Nigeria. This was followed by NASA's acceptance of the network on 1 July 1961, just a few months before the first scheduled orbital flight by an American astronaut which by this time had been put back almost a year.[18]

For the American public, however, there was one small problem. On 12 April 1961, almost three months before NASA's acceptance of its Mercury tracking network, Anna Takhtarova and her grand-daughter, Rita, were weeding a potato field near the village of Engels in the south-east of what is now Russia. Just before 11 am, they noticed an unusual object floating down towards the ground in a cluster of parachutes. By the time they reached it in a stubby field, there was a strange figure dressed in an orange space suit standing beside what looked like a very large and badly charred ball.

'Hello! I'm a friend, comrades,' he said.

'Have you really come from outer space?' asked Anna.

'Just imagine it, I certainly have,' Yuri Gagarin replied cheerily.

Gagarin had just completed the first manned space flight lasting 1 hour and 48 minutes, including a single orbit of the Earth at a speed of up to 17 500 miles per hour. Having taken over from President Eisenhower less than twelve weeks earlier, it was left to President Kennedy to explain to the American public: 'We are behind. The news will be worse before it gets better, and it will be some time before we catch up.' What most Americans didn't know was that Gagarin had safely undertaken his orbital flight without the benefit of a worldwide tracking network.[19]

Before This Decade Is Out

In the eleven weeks between his inauguration and Yuri Gagarin's orbital flight, President Kennedy had expressed serious doubts about the cost of sending a man to the Moon. 'Can't you find me something to do here on Earth that would use the money more effectively?' he asked his science adviser. But on hearing of Gagarin's triumph, a rattled President interrogated this adviser again.

> Is there any place we can catch them? What can we do? Can we go around the Moon before them? … Can we leapfrog? … When we know more, I can decide if it's worth it or not. If somebody can just tell me how to catch up. … I don't care if it's the janitor over there, if he knows how. The cost, that's what gets me. Thirty billion bucks and we don't even know if the damned thing will work![1]

Shortly afterwards, American-backed Cuban rebels landed at the Bay of Pigs, determined to overthrow Cuba's left-wing dictator, Fidel Castro. As Castro's troops battled the invaders, Kennedy wrote to the Soviet premier, Nikita Khrushchev, warning him against intervening. Although Castro soon had the rebels on the run without any help from Khrushchev, Cuba and the Soviet Union became firmly aligned. With Moscow's sphere of influence now extended to within just 90 miles of the Florida coast, the humiliated American president dogged by what had quickly been dubbed 'The Bay of Pigs Fiasco' and desperate for a propaganda win, summoned his vice president, Lyndon Johnson:

> Do we have a chance of beating the Soviets by putting a laboratory in Space, or by a trip around the Moon, or by a rocket to go to the Moon and back with a man? Is there any other Space program which promises dramatic results in which we could win?[2]

Having been associated with NASA since its earliest days, Johnson knew exactly who to consult: Wernher von Braun. 'We do not have a good chance of beating the Soviets to a manned laboratory in Space,' von Braun advised. 'But we have an excellent chance of beating…[them] to the first landing of a crew on the Moon.' Accepting this, President Kennedy addressed a joint session of Congress on 25 May 1961. With Vice President Johnson in the chair, the President said:

> I believe this nation should commit itself to achieving the goal, before this decade is out, of landing a man on the Moon and returning him safely to Earth. No single space project in this period will be more impressive to mankind, or more important for the long-range exploration of Space; and none will be so difficult or expensive to accomplish.[3]

Kennedy had been fortified in making this historic pledge by the sub-orbital flight of the first American into space, Alan Shepard. Blasting off from Cape Canaveral aboard one of von Braun's Redstone rockets on 5 May 1961, Shepard's flight had lasted just 15 minutes and 22 seconds. After reaching an altitude of 116 miles, Shepard landed in the Atlantic Ocean, 300 miles east of his launch pad. NASA's next manned flight, by Virgil 'Gus' Grissom on 21 July 1961, was also sub-orbital. It lasted only 15 seconds longer than Shepard's. And Grissom nearly drowned when his capsule sank to the bottom of the Atlantic. Within weeks, the Soviets struck again when Gherman Titov completed almost eighteen orbits over twenty-five hours. During that time, Titov showed that a man could work, eat, drink and sleep in a gravity free environment. Soon afterwards, NASA issued a media release. Despite the Soviets' obvious success with manned orbital flights, this release said: 'The men in charge of Project Mercury have insisted on orbiting a chimpanzee as a necessary preliminary checkout of the entire Mercury program before risking a human astronaut.'[4]

Although far-flung stations such as Muchea and Woomera were not involved in tracking the sub-orbital flights of Shepard and Grissom, they were being made ready for NASA's orbital missions. About fourteen days before such a mission, a team of three NASA operatives would arrive at Woomera and at every other network station. Each team comprised a capsule communicator (CapCom) to speak directly to the astronaut in space; a systems' monitor to review the capsule's status, based on its telemetry; and a flight surgeon to monitor the astronaut's health, also based on telemetry. To support these NASA operatives at Woomera for the duration of a mission, Tom Reid's team was responsible for tracking the capsule from the moment it whizzed over the horizon, feeding its telemetry to the systems monitor and surgeon, and maintaining two-way voice communication between CapCom and the astronaut for the seven minutes or so that the capsule was in view.[5]

After being given a tour of NASA's Woomera tracking station during the early days of the Mercury Program, the author Ivan Southall described Tom Reid's domain.

> The main control room with its boxes packed with electronic devices looks like a library but the bookshelves are sealed and the doors are shut except to a favoured few. These few, the engineers and technicians, endlessly maintain, endlessly check and endlessly consult these tall

> grey boxes, each to be used or to be referred to or to be triggered by an electronic mind in space. Here is a two-way radio, the finest man can build, listening to the voice of the astronaut in orbit and transmitting to the astronaut the reassuring sound of a human voice from the ground. Here, too, is a 90-channel telemetry system measuring every conceivable quantity in the Mercury capsule, and every vital physiological function in the body of the astronaut ... The control panel at which sat an astronaut [CapCom], a doctor and an engineer, were masterpieces of plain common sense. If an instrument registered the pulse rate of the astronaut in orbit, it was labelled 'Pulse Rate' and the calibrations were shaded in green for safe, in yellow for caution, and in red for danger.[6]

Mercury-Atlas 4 was NASA's first orbital flight. In place of any living thing, its payload consisted of a pilot simulator and two voice tapes to try out the worldwide tracking network, together with instruments to monitor noise, vibration and radiation. As it was a dress rehearsal for an orbital flight by a chimpanzee, a three-man NASA team was sent to each tracking station two weeks before the scheduled launch on 13 September 1961. According to the *Mercury Handbook*, the Woomera team, whose CapCom George H. Guthrie was a twenty-nine-year-old NASA aeronautical engineer, would face inconveniences. 'Recreational facilities are somewhat limited', it said, 'but a movie theatre and swimming pool are provided.' The Africa-bound teams, however, faced physical danger and the handbook warned that the local political situation was 'extremely risky'. During Mercury-Atlas 4, the Zanzibar station was surrounded by rioting local tribes. For NASA's person in charge of remote site teams, Gene Kranz, Atlas 4's one orbit mission involving the first deployment of a fully global tracking network was 'a time of trial'. But it was also a success as hundreds of people all over the world, including Tom's team, learned to work together. The way was now clear to send up a chimpanzee.[7]

If the Americans' progress seemed glacial, it was because everything NASA did was in the public eye. Unlike Yuri Gagarin who had blasted off in secret as an unknown air force pilot, America's first seven astronauts, Scott Carpenter, Gordon Cooper, John Glenn, Gus Grissom, Wally Schirra, Al Shepard and Deke Slayton had been public figures since their selection in 1959. Along with their families, each of them had been extensively profiled in *Life Magazine* as god-fearing all-American husbands and fathers who would soon hurtle into space in the service of their country. Kept firmly under wraps was NASA's expectation that one or two of them would be killed before Project Mercury had concluded. If that happened, NASA needed to be able to show that all reasonable tests had been carried out beforehand. As a dictator, Premier Khrushchev could spend billions of roubles, cut corners and risk lives in secret. But President Kennedy relied on Congress and therefore on favourable public opinion to spend much, much more on NASA. And that is why NASA sent a chimp into orbit before sending an astronaut.[8]

Having established a special 'college' to train up a large number of chimpanzees, NASA finally settled on chimp number eighty-five, who by this time had spent 158 days being tested in centrifuges, heat rooms and jet aircraft. Born in the French Cameroons, his NASA handlers named him Enos which in classical Greek means 'man'. Described by his veterinarian as 'quite a cool guy and not the performing type at all', Enos was trained to recognise the odd symbol from a group of three, in a four-test cycle combining various arrangements of circles, squares and triangles, that would be repeated again and again for the duration of his flight. If he signified recognition by pressing the correct lever, he would be rewarded with a banana-flavoured pellet. If he pressed the wrong one, he would receive a mild electric shock to his left foot. At Woomera, the three-man team sent out by NASA to work hand in glove with Tom Reid comprised Fred Volpe as CapCom, William L Wafford as Systems Monitor, and Edwin L Overholt as Flight Surgeon. Because a primate would be aboard, the flight surgeon's role loomed large. And Overholt was first class, later counting Dwight Eisenhower, Richard Nixon and Douglas Macarthur among his patients. But for now, Overholt's patient was a monkey.

After blasting off and entering Earth orbit on 29 November 1961, Enos performed well. But the telemetry coming in from Tom's team at Woomera and other network stations soon indicated problems. One of spacecraft's thrusters was propelling it out of its normal orbital mode before being automatically corrected in a cycle that kept repeating. The cooling system was also playing up and Enos's body temperature soon topped 100 degrees. Worst of all, the four-test cycle broke down when it got stuck repeating a pattern of a circle, followed by a triangle, followed by another circle. Although Enos kept responding correctly by repeatedly pressing the centre lever, it soon broke and he was subjected to seventy-nine shocks in a row. Even fifty-five years later, Bill Miller could recall the spikes in Enos's respiration, temperature and blood pressure that were recorded at Woomera. From this and from similar telemetry coming in from other network stations, together with advice from Ed Overholt and the other flight surgeons, NASA was alerted to Enos's distress. And a decision was taken to end his flight after two orbits rather than the scheduled three. Just then, a tractor outside Tucson, Arizona, accidentally ploughed up a key cable, cutting off a vital network link which was necessary to initiate an early re-entry. With only twelve seconds to spare, this key connection was restored and Enos touched down safely. Concluding that an astronaut would have been able to fix the things that had gone wrong in Enos's capsule, NASA decided the time had come for a manned orbital mission. On 29 November 1961, Enos attended a press conference for the announcement that John Glenn would be the first American astronaut to attempt an orbital flight.[9]

Unlike many of his Woomera colleagues, Tom Reid had not been carried away by President Kennedy's soaring rhetoric; nor was he enthused by what it had sparked: the superpowers' race to the Moon. As a boy Tom had predicted that a man would walk on the Moon during his lifetime but being personally involved was not the be-all and end-all for him. Not long before his death in 2010, Tom was asked whether the Apollo program had been part of a plan to bring about the demise of the Soviet Union. He replied:

> Apollo was part of a US geo-political strategy which included many other programs and it was difficult to ascertain its effectiveness in the demise of the USSR…I have serious doubts concerning the role of the US government in geopolitical activities since the early 1950s. None of my comments are intended to deprecate the skills and professional conduct of the astronauts, most of whom were highly trained test pilots, and some, good friends of mine.[10]

This did not mean that Tom was a left winger, far from it. As his younger daughter, Danae, said on his seventieth birthday: 'He's more right than Ghengis.' Rather, it meant that Tom was not ideologically wedded to NASA for the duration of the race to the Moon. And by late 1961, he had finalised plans to leave Woomera for an academic career. Not wanting to prematurely tip off his superiors, Tom bypassed Woomera's secretarial pool and prevailed upon Bill Miller to help prepare his application to the South Australian Institute of Technology. As a former police officer, Miller could type and he used his personal machine with a Gothic font to prepare Tom's paperwork. Noticing Miller's puzzled look, Tom explained that he had long harboured plans to become an academic and to undertake a PhD.[11]

Tom Reid's application to commence as a senior lecturer in electrical engineering from the beginning of the 1962 academic year was successful. He gave notice to NASA, effective from 23 December 1961. What this meant was that unless John Glenn's orbital flight took place as scheduled on 20 December, Tom would miss it altogether. As it turned out, that first launch date could not be met. While his old team continued on with a series of gruelling simulations over Christmas for the next scheduled launch on 16 January 1962, Tom had time to spend with his family which now included a fourth child, Danae, who had been born at the Woomera Hospital on 8 March 1961. As the Reid family packed for their move to Adelaide, they enjoyed one last Woomera Christmas. This included Santa's arrival which was notified in the local paper, the *Gibber Gabber*, with semi-military precision.

> Children's Christmas Party: at 2.00 pm at the School … Father Christmas will arrive by helicopter with a present for every child. Children: Please keep clear of the helicopter. There will be free train rides in a most wonderful train, with two carriages, guard's van and real engine. There will be sweets, balloons, and drinks for all the children, and they are requested not to throw away their drinking container if they want another drink. A cup of tea will be available for parents.[12]

For Tom and Betty, there was also the Christmas party to look forward to in Woomera's senior mess, where a cocktail rather than a cup of tea was the preferred beverage. With the men dressed in dinner suits or at least a suit and tie, their partners matched them in summery formal dresses of all colours, shapes and sizes. Decked out in a slip dress which accentuated her petite figure and with her dark hair swept up in a stylish bun, Betty was not unlike Audrey Hepburn to look at as Tom, suave in a dark suit, twirled her around the dance floor. They were a striking couple. For a few hours, with the mess done up in brightly coloured baubles and tinsel, Tom and Betty and everyone else there were able to forget the challenging desert environment outside. Even so, it was outdoors in a rocket park specially built beside the senior mess that Tommy, Marg and the older children of the other partygoers happily played among the various missiles on display, while the younger ones like Nick and Danae were minded by single women who were happy to be casual babysitters, when not working out on Range E.[13]

Tom Reid was pleased that the NASA team assigned to Woomera for John Glenn's flight would again comprise Fred Volpe, William Wafford and Ed Overholt. This meant that for Tom's successor there would be continuity. Having departed the United States in the late autumn for Enos's flight, with an assurance from the *Mercury Handbook* that Woomera's average summer temperature would be a mild 23°C (72.9°F), the Americans were surprised to discover, as John Glenn's flight was further delayed, that Woomera's summer temperature could spike as high as 49°C (120°F). Their main consolation was that they would be working in an air-conditioned building cool enough for Tom to have okayed his team's request for a Greek chef, who grilled up their favourite dish, souvlaki, accompanied by fried potatoes.[14]

Many years later, Tom summed up his time working for NASA at Red Lake with typical modesty.

> I became involved in Project Mercury in 1960. They had a little tracking station at Red Lake near Woomera. I provided some assistance on the telemetry side, installing, checking it out and getting it ready for the first [manned] mission.

Although Tom's tendency was to move on to his next job without regret and without looking back, he maintained an intense interest in Project Mercury, and in the performance of his old Red Lake team, devouring media reports on the progress of John Glenn's mission, as well as receiving regular updates of a more confidential nature from Fred Volpe and from Bill Miller who was to become Tom's lifelong friend.[15]

In this way, Tom shared the frustration of his old team at Red Lake as the successively scheduled launch dates of 16, 23 and 30 January came and went, due mainly to bad weather and mechanical faults on the launch pad. These were followed by further delays as one launch after another on 13, 14, 15, 16 and 18 February fell through. On each occasion, the Red Lake team had been involved in a network countdown. These began about six hours before a scheduled launch and involved checking out all Woomera's systems, together with its voice and data links to other stations. After yet another one of these, which lasted well into the evening of 20 February Woomera time, followed by numerous pauses in the launch countdown, John Glenn finally blasted off from Cape Canaveral at 9.47 am, which was about 11.17 pm in Woomera. Travelling east and then south-east around the Globe, Glenn expected to appear over Woomera approximately one hour later. Having passed straight over Perth, Glenn hurtled on towards Woomera at a speed of 17 500 miles per hour. A few minutes later, when *Friendship 7* was approximately 200 miles east of Esperance, the Red Lake team locked on to Glenn's signal and quickly established two-way voice communication.[16]

> Volpe: '*Friendship 7*, *Friendship 7*. This is Woomera CapCom. I read you loud and clear. We are standing by.
>
> Glenn: 'Roger Woomera'.
>
> Volpe: '*Friendship 7*. We have your blood pressure readout: 126 over 90. What's the result of your psychologic tests? Over'.
>
> Glenn: I have had no ill effects at all as yet from any zero G. It's very pleasant in fact. Visual acuity is still excellent. No stigmatic effects. Head movements cause no nausea or discomfort whatsoever. Over.'

This was followed by a lengthy exchange during which Glenn read out the status of his capsule's many systems, among them a cabin temperature of 100 degrees and humidity of twenty-five per cent, all faithfully recorded on Ampex tapes by Bill Miller. Having told Muchea's CapCom a few minutes earlier that he could see the lights of Perth which had been especially switched on for him, Glenn was advised by Volpe that Woomera had done the same. When asked if he could see them, Glenn replied: 'Negative. There's too much cloud over this area. I had the lights of Perth in good shape. They were very clear. But I do not have the lights of Woomera, sorry. Over.' Meanwhile the nearby FPS-16 Radar which was slaved to Red Lake's telemetry antennae, had zeroed in on Glenn's precise location. And every six seconds it sent updated location data to Goddard. Less than nine minutes after *Friendship 7*'s signals had first been picked up by Woomera, it passed over the horizon and beyond view. All in all, the Red Lake team had performed well during Glenn's first pass, maintaining a steady telemetry feed, an excellent two-way voice link and precise location data. Goddard had been passing this data on to the next station at Canton Island, enabling its antennas to be pointed so they could lock on to *Friendship 7* at the first possible moment.[17]

As Glenn hurtled on over that station towards California, Mission Control received a call from the White House; President Kennedy wanted to speak to Glenn at the end of his first orbit. Noticing the irritated look on his network controller's face, the flight director, Chris Kraft, barked: 'Make sure everything is set up … The president is the boss!' Within minutes, Glenn was over Cape Canaveral whose CapCom, Alan Shepard, called him up:

> Shepard: '*Seven*, this is the Cape. The President will be talking to you.'
>
> Glenn (stammering): 'Ah……..the President? This is *Friendship 7* standing by.'
>
> Shepard: 'Go ahead, Mr President'.

But there was no reply because the President's call had come early and the White House line had not yet been patched in to NASA's communications loop. George Metcalf, the Cape technician who had picked up the phone, thought it was a gag when he first heard the voice at the other end say: 'This is the White House. Stand by for the President.' But then on hearing that famous Boston accent, Metcalf stammered: 'Hello, hello, Mr President!' As Metcalf gestured wildly to get a patch fixed up, Chris Kraft was hit with the news of a segment 51 warning which suggested that the capsule's heatshield and compressed landing bag were no longer locked in position. 'Forget the President's phone call,' Kraft said testily, 'and verify the patching of segment 51'. It was left to Metcalf to deal with the White House call. 'Mr President, we've gotten pretty busy down here now,' Metcalf said. 'I don't think we've got time to talk.' To this the President replied: 'Give me a call if you get a chance.'[18]

Meanwhile Woomera and all the other network stations had been ordered to closely monitor the telemetry relating to segment 51. Volpe and his fellow CapComs were told to let Glenn know that the landing bag deploy switch should be in the 'off' position; nothing more. After hearing this a couple of times, Glenn cottoned on to what his controllers at Cape Canaveral were reluctant to spell out to him: that he might be at risk of burning up during re-entry. Nevertheless, as he spoke to Volpe while passing over Woomera on his second orbit at about 1.45 am local time, he sounded remarkably calm, with his 30-minute status and medical reports indicating that he was holding up very well. And on his third and final pass over Woomera at about 3.15 am, Glenn felt relaxed enough to yaw his capsule, turning it around 180 degrees so he could watch his last sunrise in space head on.[19]

Chris Kraft had a hunch that the segment 51 readout might be false telemetry from the capsule. Even so, it was decided that after Glenn had fired his retro rockets triggering the final re-entry phase, the spent retro package should be retained to keep the heat shield in place just in case it had come loose as the read-out suggested. But all went well. And a hot, sweaty and fatigued Glenn emerged from his capsule after it had been plucked from the Atlantic Ocean by the destroyer USS *Noa*. During Glenn's post-flight medical check-up, it was discovered that he had lost well over five pounds in weight during his five-hour flight.[20]

A few days later, President Kennedy presented John Glenn with NASA's Distinguished Service Medal. Also present were the Glenn and Kennedy families. When the President's four-year-old daughter, Caroline, was introduced to Glenn she asked, 'Where's the monkey?'[21]

Academic Interlude

Formed at the beginning of 1960, the South Australian Institute of Technology incorporated the School of Mines and Industries which boasted alumni such as the legendary industrialist, Essington Lewis. Prior to its incorporation into the new institute, this school's focus had been on technical education at all levels, including a two-year trade course in television to train up technicians and studio operators. One of its students, Ed von Renouard, would later play a crucial role at Honeysuckle Creek in the live televising of Neil Armstrong's first step. The express purpose of the new institute was to specialise in applied science and technology at the tertiary level. The South Australian Premier, Thomas Playford, hoped that it would cater to such of the technological and industrial needs of his state which were not being addressed by the University of Adelaide. He didn't want it to evolve into a more generalist university like the New South Wales University of Technology. To that end, a division of the institute was established in the steel-making city of Whyalla and a number of subjects were also taught at Woomera, including Engineering Drawing and Design I.[1]

The institute's main campus, a site of almost five acres on the corner of North Terrace and Frome Road, Adelaide, had as its centrepiece a handsome five-storey building in the Gothic style. It was here that the electrical engineering and electronic engineering departments were located. Each of them offered degrees incorporating innovative courses, including one for businessmen and engineers which covered the principles of both digital and analogue computers. In addition to its links to Woomera, the institute had a close association with the Weapons Research Institute.[2]

Apart from the prospect of undertaking a higher degree himself, what mattered most to Tom Reid as one of the institute's senior lecturers was the opportunity to inspire students to think innovatively about the ever-accelerating rate of change in technology, especially computing, that they would face during their working lives. In lecturing and examining undergraduates in communications theory, pulse techniques and industrial electronic courses, Tom was passionate about encouraging well-grounded flexible thinking. 'My general approach,' he said, 'was to offer a sound basic foundation to minimise the possibility of early obsolescence resulting from changes to technology.'[3]

During Tom's time at the institute's electrical engineering department, an average of fifty students per year were enrolled in degree courses, making it slightly larger than its Glasgow counterpart during Tom's own undergraduate days. As with Glasgow, the institute's electrical engineering courses were taught and examined with great rigour, often resulting in a combined wastage and failure rate of seventy-five per cent. Even so, many undergraduates who could not pass their degree courses still benefitted from what they had been taught and went on to have solid careers in technical services.[4]

For Betty and her children, Tom's new job meant a move to the still developing suburb of Elizabeth Grove on Adelaide's northern outskirts, not far from the Weapons Research Establishment (WRE) at Salisbury. Tom and Betty had a number of friends there due to the WRE's constant rotation of scientific staff with Woomera while young Tommy and Marg were made welcome at the local primary school. Although this part of the city had a much drier climate than central Adelaide, it was paradise compared to Woomera. But for its artificial water supply, Woomera would have been a treeless desert, whereas in and around Elizabeth Grove, there was a natural abundance of tall, shady eucalypts. Without using too much tap water, Betty was able to grow a variety of colourful shrubs and roses in the garden beds that ringed their yellow-painted two-storey house. All the while, Tom managed to keep their front lawn green, an unimaginable luxury in Woomera. It was also a treat to be able to make regular trips to the seaside, the family favourite being West Beach not far from Adelaide airport. Tom's commute to the institute in central Adelaide on well-made bitumen roads was far less challenging than the daily grind of negotiating the long and pot-holed gravel track that linked Woomera Village to the Red Lake telemetry building. Unlike most of the jobs he'd had at Woomera and in the navy, Tom was able to keep regular hours as an academic. Life was good.[5]

All the while, the space race between the United States and the Soviet Union had continued. By the end of Project Mercury in June 1963, the Americans were hailing it as 'one of the greatest medical experiments of all time', which demonstrated that prolonged weightlessness posed no threat to astronauts' health. But the Soviets had figured this out too. And they were still ahead both technologically and symbolically when between 16 and 19 June 1963, the world's first female cosmonaut, Valentina Tereshkova, orbited the Earth forty-eight times, remaining in space seventeen hours longer than all the American astronauts put together. NASA was therefore eager to commence its next phase, Project Gemini, during which two-seater versions of the Mercury capsule would be used to change orbits, and to meet and dock. These capsules would also enable astronauts to step outside for so-called space walks. For this phase, Woomera would continue to play a vital tracking role.[6]

Although the race to the Moon captured the public's imagination, it was just one part of a broader space race. This contest included the exploration of deep space, beginning with the Earth's neighbouring planets, before moving out beyond the solar system. It also encompassed the development of unmanned Earth-orbiting satellites to be used for defence, weather-forecasting and communications. As the pace of the space race intensified, NASA came to better appreciate the importance of Australia in its global tracking network. If the early stages of a rocket flight went as planned, its satellite would pass over Australia on its first orbit. The Australian tracking stations were especially important in helping to determine whether a launch had been a success.[7]

Since its inception in 1958, NASA had developed an excellent working relationship with the Commonwealth Department of Supply. When the space agency requested the Australian government's agreement to a new site for its latest planned tracking stations, preferably in the south-east of the continent, Prime Minister Robert Menzies was only too happy to oblige. NASA's preferred location had to be reasonably close to a city with good communication links, industrial facilities and a university. It also had to be an attractive place to live. There were numerous stories of families flying in to remote outback settlements built to support tracking stations who would decide to return to Britain before their aircraft's propellers had stopped spinning after seeing the arid landscape during their plane's descent, which was enough to change their minds. Another requirement was a location remote from radio interference and heavy air traffic. As it turned out, the Prime Minister's pet project, Canberra the national capital, looked promising. It was serviced by Fairbairn Airport. It had an industrial area in Fyshwick. And it was home to the Australian National University. Soon surveyors, geologists and engineers were fanning out into the mountains just south of Canberra looking for suitable tracking station locations. Some of the country was so rugged that they had to use helicopters.[8]

In terms of its building program, NASA had three priorities: a station to be part of its deep space network (DSN); another to support its satellite tracking and data acquisition network (STADAN); and a third to form part of its manned space flight network (MSFN). Focusing initially on the first two, NASA settled on a site not far from the Murrumbidgee River for its deep space tracking station. Known as Tidbinbilla, this site was close to Canberra but separated from its radio waves by the Bullen Range. Due south of Tidbinbilla was a second range beyond which stretched the magnificent Orroral Valley, about 3300 feet above sea level. Just as Tidbinbilla was shielded from Canberra's radio waves by the first range, the Orroral Valley was shielded from Tidbinbilla's radio waves by the second range. The ancient Paleozoic rock formations delivered up an ideal environment for space tracking at these two sites.[9]

By March 1963, NASA and the Australian government had agreed to lease 150 acres of prime cattle grazing country in the Tidbinbilla Valley to build an 85-foot diameter dish and its associated deep space tracking equipment. Now the race was on to complete enough of the work in time to support Mariner 4, NASA's first probe to Mars. A construction contract was let to AV Jennings on 1 July 1963 and RA (Bob) Leslie was appointed Tidbinbilla's first director at around the same time. After graduating with honours in electrical engineering from the University of Melbourne in 1947, Leslie had worked on pilotless aircraft in both Britain and Australia before commencing a long association with NASA. He was considered a father figure of Australia's tracking stations. With work on the Tidbinbilla deep space tracking complex well underway, NASA's attention turned to the construction of a satellite tracking station in the Orroral Valley.[10]

Following site selection and planning, a building contract for Orroral Valley was let in August 1964, this time to the Canberra construction company, TH O'Connor. All the while, the Supply Department had been searching for a suitable person to be the first director of this new tracking station. In accordance with Australian government policy, the successful candidate would have to be an Australian citizen or resident. Heading up the search was none other than MS Kirkpatrick, who in 1963, had been promoted out of Woomera to be superintendent of the Supply Department's American projects division in Canberra. While the dish to be built at Orroral Valley would look the same as the one at Tidbinbilla, the demands on its staff would be entirely different. For one thing, the orbits of the satellites it would be tracking, involving as many as forty passes a day, could vary from close Earth orbits to highly elliptical ones, out beyond the Moon. After each pass, sometimes less than ten minutes apart, the tracking equipment would need to be completely switched around and re-configured. It would be one of the busiest tracking stations in the world, equipped with more tracking systems than any other Australian station. While the average deep space tracking station received data from spacecraft at the rate of 100 bits per second, Orroral Valley would be required to process spacecraft data at the rate of 128 000 bits per second.[11]

Although construction would be the responsibility of TH O'Connor and EMI Electronics would be tasked with providing the station's operations and maintenance personnel, MS Kirkpatrick knew that the person chosen as director would, as the public face of the station, shoulder much of the burden if things went wrong. And it would be up to the director to meld the operations and maintenance teams into an effective unit. As he mulled over his options, Kirkpatrick's thoughts soon turned to Tom Reid who, to his great disappointment, had resigned from Woomera to become an academic. Reid's qualities which had led Kirkpatrick to select him for Woomera were exactly what he was looking for but when Tom was first approached by Kirkpatrick about the Orroral Valley directorship, he was reluctant to take it up. Although he was confident of his ability and was a can-do optimist by nature, he wondered whether the demands of Orroral might be more than he could handle. Besides his family seemed happily settled in Adelaide. On the other hand, things were not going quite as he had planned at the South Australian Institute of Technology. His hopes of eventually becoming a professor had been dashed when the Institute's board decided against the introduction of this title, preferring that of 'assistant director' instead. Tom thought some more about Orroral Valley. As he and Betty discussed what living in Canberra might be like, he came to realise that his wife was upbeat about a city that promised a hint of

Scotland with its mountain views, its four distinct seasons and the possibility of winter snow. After weighing everything up, Tom signed on as the inaugural director of NASA's Orroral Valley satellite tracking station agreeing to commence in August 1964, the same month that the station's construction contract was let to TH O'Connor.[12]

Betty Reid was also attracted to Canberra because she would be closer to her own family. A younger brother, Jim, had migrated to New Zealand in 1961. Three years later, their widowed mother, Margaret McKenna, decided to resettle there too. In late June 1964, Margaret and the younger members of her family boarded the SS *Himalaya* for the long voyage out, stopping at Adelaide along the way. They were there for just one memorable day during which Betty was able to catch up with her mother, who she hadn't seen in almost twelve years. Margaret was also able to spend time with her Reid grandchildren, three of whom she had never met. Among those present was Betty's youngest brother, John, who was about to turn thirteen. 'We were met by Tom and Betty and the kids in a Holden and we all went to their place for the afternoon,' John McKenna later recalled. 'We thought that we were going to be seeing a lot more of Betty.' They all looked forward to this now that the Tasman Sea rather than half a world would be the only thing separating them. There was just one sour note; the camera John had used to take photos of this reunion was later stolen, along with its precious film.[13]

Just weeks after Margaret McKenna and her brood steamed out of Port Adelaide, bound for New Zealand, Tom left for Canberra, by then a city of approximately 75 000 souls with an explosive annual growth rate of twelve per cent. This growth was fuelled by the ongoing relocation of tens thousands of public servants from Melbourne. The national capital's housing shortage was so acute that many of them ended up staying in hostels, bachelor flats and hotels. Tom was employed by the Department of Supply in a very senior role that directly impacted upon relations between Australia and the United States. And he could have pulled rank by having someone else obtain priority housing for him. But he resisted this temptation partly because EMI Electronics had the following clause in the employment contracts of those who would be working for him: 'It must be understood that the matter of housing for yourself and family is basically your own private problem.' Determined to set an example, Tom went in search of a suitable Canberra home while the rest of the family remained in Elizabeth Grove, allowing Tommy, Marg and Nick to finish their school year in Adelaide.[14]

Whatever time Tom had set aside to go house hunting was soon cut short by a high-level announcement. On 3 September 1964, the Minister for Works, Senator John Gorton, advised Parliament that in relation to Orroral Valley, a contract worth well over 500 000 pounds had been let for the construction of an 85-foot dish, an operations building, a powerhouse and related drainage, water supply and sewerage facilities. Almost as an afterthought, Gorton mentioned a further contract worth more than 75 000 pounds for road access to the station. Although this seemed an enormous sum, Tom had driven to the site along the track used by local farmers. What he had seen made him wonder whether this budget would be enough. Adding to the pressure was a statement by Gorton that he expected completion by May 1965.[15]

Further complicating Tom's task was an earlier promise made by the Minister for the Interior, Doug Anthony, who had assured Parliament that construction of the tracking station would not affect the surrounding area which had been reserved for a future national park. This area was frequented by wallabies, rosellas, cockatoos, lizards, lyre birds, sugar gliders, koalas and echidnas, together with black snakes, brown snakes and tiger snakes. When a visiting journalist asked a local farmer whether these snakes were poisonous, he replied, 'It depends on how much Bundy [rum] you've got in your system.' Some of the local wildlife adapted remarkably well. A worker who had been standing near the access track to the station's construction site stepped back to let an approaching vehicle pass:

> I could hear this jeep coming up the road. The driver was swopping cogs (changing gears) as it came up the hills, then dropping her back as he went down again. I waited for this jeep to pass but still it came on. After five minutes, I started down the track and saw the jeep...a bloody lyrebird.[16]

The Orroral Valley construction site was located over 40 miles south of Canberra. Its existing dirt access road, a track really, was suitable for light vehicles only, especially where it wound its way up and down a steep hill, known as Fitz's Hill and crossed two rivers: Rocky Crossing on the Naas River; and Orroral Crossing on the Orroral River. Thanks to Gorton's road-building grant, the most hazardous sections including a steep and winding climb along the side of Fitz's Hill, were sealed; and both river crossings were rebuilt. But there was not enough money left to lay bitumen over the remainder or to widen the single-lane wooden-truss bridge across the Murrumbidgee River at Tharwa which dated from the late nineteenth century. No matter how well compacted and graded, large sections of the unsealed access road were reduced to impassable bogs after heavy rain while the upgraded river crossings were submerged in raging torrents of water.[17]

The first section of the Orroral Valley antenna, a massive twenty-four-ton, sixty-foot segment of the dish, the first of a series weighing about 400 tons in total, arrived in Canberra from the United States on 23 October 1964. Due to heavy rain along the access road, the massive semi-articulated low loader used to transport it, could not commence its journey because of concerns that it might get hopelessly bogged. This first bit of the dish remained stored in the Department of Supply's Fyshwick warehouse until the road dried out. Finally, on 4 November 1964, the all clear was given. But as the low loader approached the Tharwa River Bridge it ground to a halt. No one was sure whether the dish segment could be squeezed between the bridge's ancient wooden trusses.[18]

Castles in the Air

The surveyors had calculated precisely. And in late 1964, load after load of antenna parts were successfully carried across the Tharwa Bridge on massive low-slung trailers. The bridge's carriageway was made up of heavy pieces of lumber on top of which ran two parallel lines of sturdy timber planks. Normally even fully laden cattle trucks were safe if their wheels remained somewhere on these planks. But not so the trailers carrying antenna parts. To ensure that they didn't bring the bridge's ancient wooden trusses crashing down, each prime mover was preceded by two men, slowly walking backwards. Their job was to use hand signals to indicate to the driver how much room he had on each side, sometimes just fractions of an inch. For those involved, the large signs fixed to the rear of each trailer, 'Danger Wide Load', seemed like an unnecessary afterthought.[1]

Once all the antenna segments and other structural materials had arrived on site, construction of the 85-foot dish and its related buildings proceeded quickly. Much had been learned from the delays and cost over-runs at Tidbinbilla. By May 1965, Orroral Valley's essential infrastructure had been completed ahead of time and under budget. The challenging part was to put together the operations building's innards: all the cabling, computers, control panels and other bits and pieces necessary to drive the dish, to allow multiple satellites to be tracked in one day, to make sense of their downloaded data and to package it all up for electronic transmission back to the United States. In this, Tidbinbilla provided no precedent because as a deep space tracking station, its transmission bit rate was just one thousandth of that required at Orroral Valley to track Earth orbiting satellites. It followed that the innards of the new station were infinitely more complex than those at Tidbinbilla. Above the solid floor of Orroral's main station building, a networked series of supporting frames were constructed upon which a further floor of large moveable tiles was placed. The space between these two floors created a void or 'plenum' throughout which a complex network of cables and ducting was laid. These linked up to all manner of equipment on the tiled floor above including computers, recorders and receivers, and also provided air-conditioning to keep the station's myriad electrical circuits from overheating. Their complexity sorely tested Tom Reid and his team, as well as their American advisers. 'It was just total confusion,' recalled one engineer, Alec Hempstead. 'There were cables piled up everywhere, equipment laying here, equipment laying there, and people just milling around.' Then after a short time, things started to take shape and the operations room began to look ready for business. According to Hempstead:

> It seemed to occur very suddenly, after about a fortnight's work moving things around [and] occasionally watching the poor Americans put the dish into the ground. You would hear a mad scuffle, as they made changes to the servo system, then the dish was heading to the ground and there would be a cry of OW! STOP! They put the dish into the ground about twice while I was there.[2]

All the while, Betty Reid had remained living in Elizabeth Grove to allow Tommy, Marg and Nick to finish their school year in South Australia. But Betty's hopes of seeing more of her mother in New Zealand were dashed at the end of 1964 when she received news that her family, with the exception of her brother Jim, was returning to Scotland after just six months. Margaret McKenna was desperately homesick for Glasgow. Money was tight and she could not afford a return voyage that included a call at an Australian port. Instead, she and the rest of her family took the most direct route back to Britain, sailing north-east from Auckland via the Panama Canal. As Betty later wrote: 'We all built castles in the air about going to New Zealand.' She would never see her mother again.[3]

With Betty and the children due to arrive in Canberra for the beginning of the 1965 school year, Tom had managed to find some accommodation – a duplex in Hackett, one of Canberra's northern suburbs, predominantly populated by clerks and tradies. Although this place in Antill Street was small for the family and a long way from the entrance road to Orroral Valley, it was the best he could hope for in Canberra's tight real estate market. Betty made do with what she called her 'government house', getting wall-to-wall carpet laid in the living room and making curtains for the children's smallish bedrooms. She wrote to her mother that while the house was a squeeze, there were many compensations. Canberra was 'the most beautiful city in Australia'; the local schools were excellent; there was a 'most magnificent convent' down the road; and the autumn leaves and surrounding mountains brought back memories of Scotland. But best of all and most unusually for Canberra, deep snow fell in the suburbs during the winter of 1965. 'I have seen my first snow since I left home,' Betty said, 'and the kids were wildly excited.'[4]

Through the winter of 1965, Tom and his team continued the complex task of getting Orroral Valley ready for tracking. This was scheduled to begin in early October. While some of the team's methods were unorthodox, they were effective. It was vitally important that the cables which fed signals to the dish were kept dry and uncontaminated. To allow for this, they were pressurised with nitrogen. But after being installed, some leaked. Soapy water was then applied to all the cable joints and wherever this solution bubbled, a leak was identified. Apart from challenges at the station, there were often problems just getting there, particularly with kangaroos. On one occasion, as a station car was going up Fitz's Hill, a kangaroo jumped onto the vehicle's bonnet before jumping off again. When one of the passengers, a new recruit, asked: 'Does this happen often?' the driver replied, 'Oh yes, all the time.' Writing to her mother, Betty Reid described how Tom was faring as director.

> He is settling into his little kingdom in Orroral Valley and loving every minute of it. This job could have been tailor made for him and he revels in the authority. He meets a variety of interesting people and has to make big decisions but he has just the capabilities for this type of thing. It went to his head a bit at the beginning but he has settled down and now drives a lovely '65 Valiant. Fully automatic, it is a dream to drive.[5]

Even before Betty had left Elizabeth Grove, she had been suffering from recurring ill health, the cause of which her doctors never seemed able to precisely identify. The most intractable symptom was extreme fatigue. At first this was ascribed to depression and treated as such. 'I thought I was dying,' she wrote to her mother. 'It was an effort to get my breath.' After arriving in Canberra, Betty went to a new doctor. After discovering that her blood count was down, he put her on a course of iron supplements. This seemed to do the trick. 'I'm feeling better than I have for years,' she said. It helped that Tommy, Marg and Nick were doing well at their new Canberra schools, allowing Betty to spend more time with four-year-old Danae. Even though their duplex was small, Tom would often bring friends, work colleagues and American contractors home for drinks. These were informal occasions enjoyed by the whole family, with Betty dressed in her casual Audrey Hepburn style.[6]

Early springtime in Canberra is sublimely beautiful as all sorts of blossoms come into flower and people emerge from their winter hibernation to enjoy deep blue skies and a warming sun.

On 4 September, the *Canberra Times* ran a prominent page three story highlighting the fact that Orroral would be a prime station, tracking a new generation of 'observatory' satellites. 'It would be capable of handling three satellites at once,' the article said, 'transmitting commands and receiving information from all of them.' In consultation with NASA, 1 October had been fixed as the day upon which the station's tracking operations would begin. It was NASA policy for staffing to be organised by private contractors and, in the case of Orroral Valley, by EMI Electronics. But as the first day of operations approached, EMI had still not been able to fill all positions. This forced the Department of Supply to put a blanket ban on technical staff taking recreation leave. And it was Tom Reid's responsibility as director to make the station work effectively from day one despite these festering problems.[7]

Not surprisingly, Tom was up early on Friday, 1 October. It was the most important day of his professional life. If he needed any reminder, it was splashed all over page one of that morning's *Canberra Times*, together with two huge photos. Under the heading 'New eye in the sky', the article explained that Orroral Valley was a prime unit in NASA's tracking network and that the station would make twenty contacts with satellites 'if all went well'. This was a reference to the fact that Orroral was still 'a maze of exposed wires and partly assembled equipment'. The minister for supply had something to say, as did Tom. 'Initial results of tests with aircraft,' Reid told the paper, 'had shown the station was performing better than the ones in North Carolina and Alaska.' The pressure was on.[8]

Betty wished Tom well for his big day. And then he set off for the station, his mind focused on the challenges ahead. Unlike the staff at Tidbinbilla's deep space tracking station, Tom's team had not been given any tracking training in the United States. Whatever instruction they received had come from those staff members who had picked up tips from the American installers and from Tom, who just over three years earlier had tracked Enos the monkey in Earth orbit. Although mistakes were made that first

With his father and grandfather standing behind, Thomas Reid the seventh sits on his great-grandfather's lap, 1927. (Courtesy Hon. Margaret Reid AO)

Petty Officer Tom Reid RN in Singapore, 1946. (Courtesy Hon. Margaret Reid AO)

Betty Reid with Thomas Reid the eighth at HMAS *Cerberus*, south-east of Melbourne, 1953. (Courtesy Marg Reid)

The Project Mercury telemetry building in the desert near Woomera. From there, Tom Reid's team tracked Enos the monkey during his orbital flight in 1961. (Courtesy Bill Miller and Colin Mackellar)

Addressing a joint session of Congress on 25 May 1961, President John F Kennedy commits to 'achieving the goal, before this decade is out, of landing a man on the Moon and returning him safely to the Earth'. (Courtesy NASA, Wikimedia Commons, public domain)

Tom Reid (left) with Department of Supply staff. Behind them is a list of antenna parts for the Orroral Valley dish, 1965. PAA refers to Pan American Airways. (Courtesy Colin Mackellar)

Tom and Margaret Reid on their wedding day, 25 February 1967, with Tommy, Nick, Danae and Marg. (Courtesy Hon. Margaret Reid AO)

Tracking station staff crossing a flooded Orroral River, no date. (Courtesy Jim Thompson and Colin Mackellar)

A station car negotiates a washaway on the old Apollo Road, leading down from Honeysuckle Creek, 1967. (Courtesy Hamish Lindsay and Colin Mackellar)

With assistance from Honeysuckle's pet kangaroo, Tom Reid presents a pair of sandals mounted on a plaque to NASA's simulations team leader: 'To Mr G. Harris Jr in the hope that his feet will be dry when walking on water', 29 September 1967. (Photo Hamish Lindsay. Courtesy George Harris Jr and Colin Mackellar)

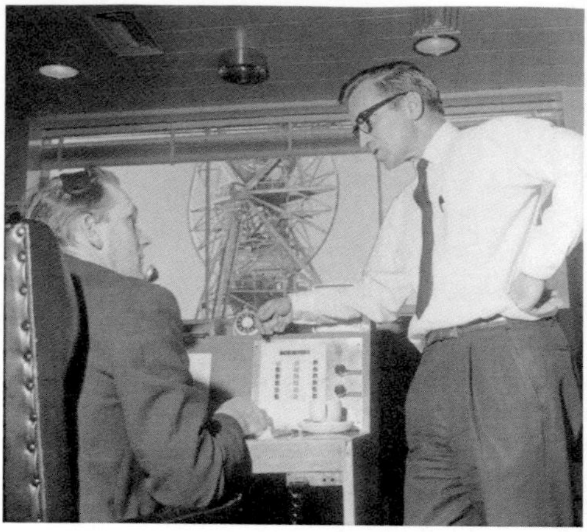

Tough and competent, Tom Reid issues directions to Honeysuckle's servo operator, Brian Bell. Visible through the window is the dish's support structure. (Courtesy Colin Mackellar)

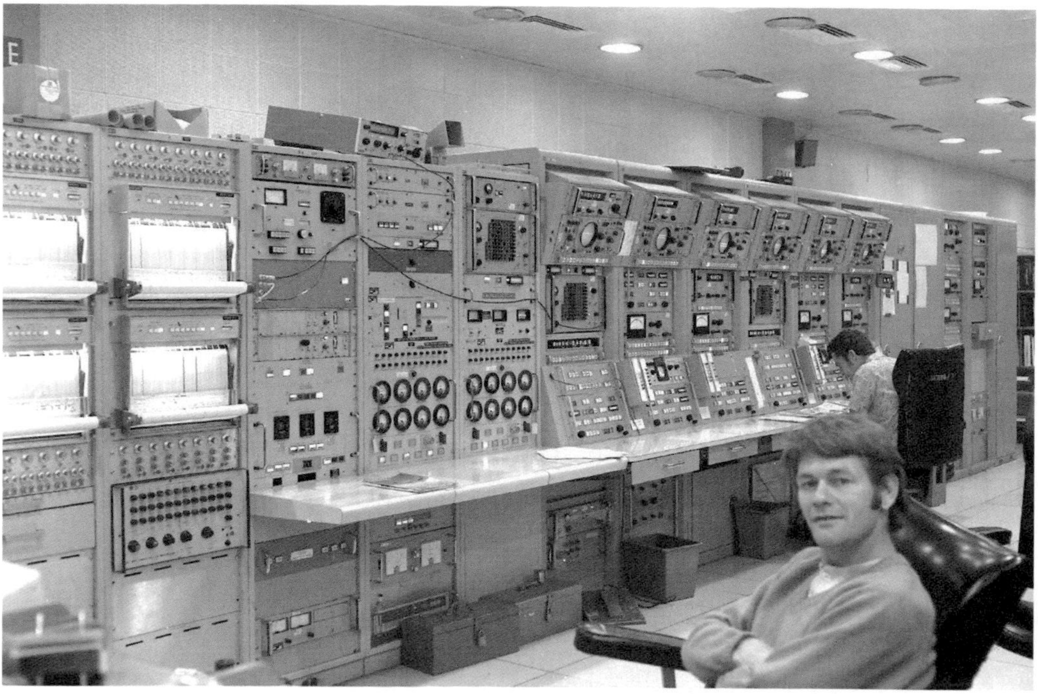

Kevin Gallegos (foreground) in front of a bank of equipment that sent and received signals passing backwards and forwards between the Honeysuckle Creek tracking station and Apollo astronauts on the Moon. (Courtesy Hamish Lindsay and Colin Mackellar)

The Honeysuckle Creek tracking station complex is dusted in snow a few days before Apollo 11's launch in July 1969. The operations building is to the left of the dish. (Courtesy Hamish Lindsay and Colin Mackellar)

Less than 4 hours before Neil Armstrong's Moon walk, Tom Reid does his best to explain the intricacies of the Operations Control Room to Prime Minister Gorton (far left) while John Saxon (foreground) struggles to contain his annoyance at this unannounced VIP visit. (Courtesy Hamish Lindsay Colin Mackellar)

The Prime Minister (second left) and Tom Reid (right) pose in front of the Honeysuckle dish, which has been moved out of its tracking alignment to facilitate this photo opportunity. (Courtesy Ken Sheridan and Colin Mackellar)

Because of a lack of space, Stan Lebar's TV camera had to be mounted upside down in Apollo 11's lunar module stowage bay. (Public domain).

Ed von Renouard at Honeysuckle's TV video console during Apollo 11. The little switch he used to flip the upside-down TV image right side up is above his head and marked for identification by two Dymo labels. (Courtesy Colin Mackellar)

Just a second or two before Neil Armstrong stepped onto the Moon, Hamish Lindsay snapped this iconic shot of the Honeysuckle dish, which was already transmitting the live TV images it was receiving from the Lunar Module, *Eagle*. (Courtesy Hamish Lindsay and Colin Mackellar)

A small part of the huge crowd in Central Park, New York, watches Honeysuckle's TV footage of Neil Armstrong stepping onto the Moon, their bitter divisions over the war in Vietnam forgotten in a moment of wonder and pure joy. (Public domain)

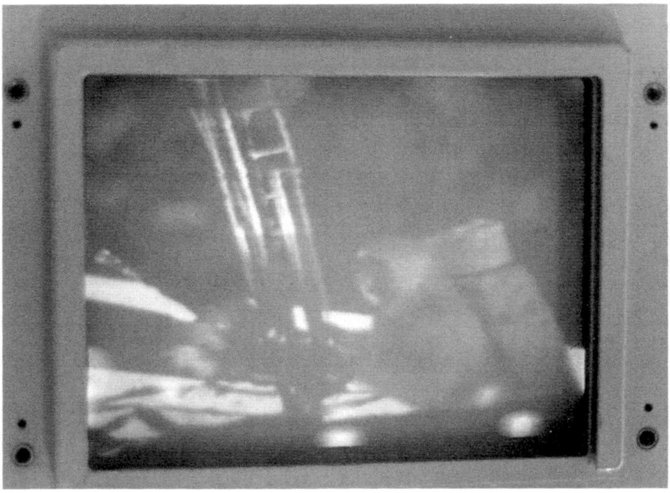

Ed von Renouard's TV monitor showing Neil Armstrong immediately after he had stepped onto the Moon. This footage was transmitted by Honeysuckle to a worldwide TV audience of 600 million people. (Courtesy John Saxon and Colin Mackellar)

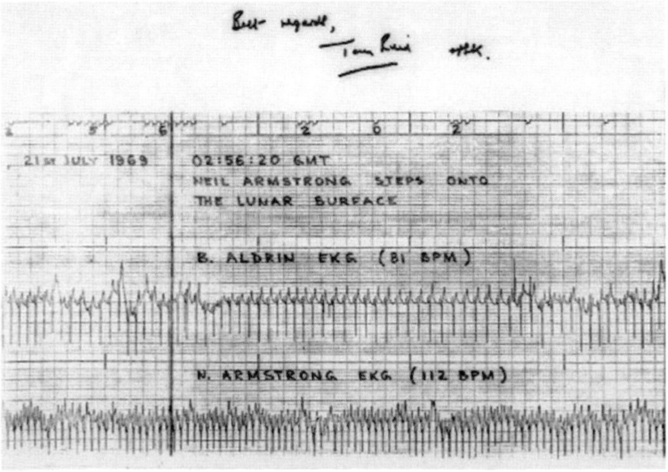

The ultimate in telemetry: Tom Reid's printout of Neil Armstrong's heart rate the moment Armstrong stepped onto the Moon. (Courtesy Colin Mackellar)

Tom Reid and the Honeysuckle Creek team he led during the Apollo 11 Moon mission. Reid is on the far left. (Photo Hamish Lindsay. Courtesy Colin Mackellar)

Pictured from left to right during the United States' president's visit to the Australian Parliament in September 1996 are: House Speaker Bob Halverson, President Clinton, Prime Minister Howard, Tom Reid, Senate President Margaret Reid, Mrs Halverson (partly obscured), Hillary Clinton and Janette Howard (partly obscured). President Clinton is signing the Senate's VIP guest book. (Courtesy Auspic)

day, Tom was reasonably happy with the station's performance. And when late in the afternoon, a *Canberra Times* reporter rang for a Saturday follow up, Tom told him:

> The changeover [of tracking from the preceding station] was not as successful as I would have hoped, but I am satisfied. It was not 100 per cent because the staff is not experienced and the satellite orbits were not suitable. But under the conditions, everything is going according to plan. During the first 24 hours of operation which ends at 10 am tomorrow [Saturday], the station hopes to pick up and track 20 satellites.[9]

Tom had many loose ends to tie up before he could hand over the evening shift to his deputy. He rang Betty to let her know that he wouldn't be home much before 9 pm. As he began the long drive back to Hackett, sometime around 7.45 pm, he looked forward to a late dinner when he could tell his wife about his first day's tracking. After winding back along the station's access road, across the Tharwa Bridge, up the Tuggeranong Valley and through most of Canberra, Tom finally turned into Antill Street. But as he approached his house, he noticed that all the lights were on. When he stepped out of his car, one of his neighbours came running to meet him, saying: 'Betty's been rushed to hospital and she's in a bad way.' Momentarily incredulous at hearing this news and deeply upset by what it might mean, Tom quickly regained his usual composure. After doing his best to comfort his children, Tom gratefully accepted his neighbour's offer to mind them overnight. He then got back into his car and raced off to Canberra Hospital.[10]

Arriving at the hospital, Tom was ushered in to see Dr Marcus Faunce, one of Canberra's leading physicians. Dr Faunce gently explained to Tom that Betty had suffered a subarachnoid haemorrhage – a stroke – and that her condition was critical. While immediate neurosurgery would normally have been attempted,

Betty had been assessed as 'not fit to move at present'. She was deeply unconscious and her eyes were fixed and dilated. Her blood pressure had fallen dangerously low and had remained that way. Distraught, Tom stayed at the hospital for some hours. After being told that his wife's condition was unlikely to change for the next little while and after signing papers to authorise surgery if the opportunity arose, he drove home. By this time his children were awake, as were his neighbours. He was then able to piece together what had happened.[11]

After eating their evening meals, the children had been in the living room while Betty prepared Tom's dinner in the kitchen. Just after 8 pm, they had heard her call out 'Tommy'. This was followed by a loud thud. Bounding out of his chair to see what had happened, twelve-year-old Tommy found his mother lying on the floor. Unable to rouse her, he had had the presence of mind to ring for an ambulance, telling the dispatcher that his mother had collapsed and that there was something very wrong with her breathing and appearance. Under lights and siren, the ambulance had arrived at around 8.15 pm. After quickly assessing Betty's condition, the paramedics loaded her into the back of the ambulance and headed off at high speed. In the meantime, Tommy, Marg, Nick and Danae had gone next door for help. When the ambulance sped away they had stayed with the neighbours.[12]

Because Canberra had long been tagged as a public service town and derisively described as 'a good sheep station spoiled', most of its residents were proud that their city had been chosen to host the largest complex of tracking stations in the Southern Hemisphere. This was a new, exciting and futuristic industry. And Canberrans followed what was going on at Tidbinbilla and Orroral Valley with great interest; hence Friday's page one story. As they read Saturday's follow-up, they would have been pleased with Tom's comments: a mixture of cautious optimism, tempered by admissions about what could be done better. In a national capital already awash with 'pollie speak', Tom's candour was refreshing. Little could they have imagined the personal tragedy that had engulfed Tom and his family just a couple of hours after he had spoken to the *Canberra Times* reporter.

Sometime early on Saturday morning, Tom contacted his deputy, Lou Cotrell, and swore him to secrecy before explaining what had happened to Betty. It was agreed that Lou would take charge of the station for the second day's tracking, allowing Tom to be with Betty and his children. That settled, Tom spent the rest of the day travelling backwards and forwards between Canberra Hospital and Hackett. As evening approached, he elected to stay by Betty's side. By now his wife was deeply unconscious and her condition was described as 'extremely poor'. At 9.15 pm she was placed on artificial respiration. The doctors described Betty's prognosis as 'hopeless'. And when Dr Faunce advised Tom of this using gentler language, he agreed to her life support being disconnected. Betty Reid passed away peacefully at 10.30 pm on Saturday 2 October 1965. She was just thirty-six years old.[13]

For Tom, the next couple of days passed in a smudged blur of disbelief. While the notice which appeared in the *Canberra Times* on 4 October laid out matter-of-factly what would happen next, no words could describe the deep and intense grief that engulfed thirty-eight-year-old Tom and his four children, none of them yet teenagers.

> **REID:** The relatives and friends of the late
> Mrs Elizabeth Reid of 268 Antill Street, Hackett are
> respectfully informed that her funeral will take place
> tomorrow, Tuesday morning. The cortege is appointed
> to leave Blackfriar's Church, Watson, after prayers
> commencing at 10 am for the Canberra Cemetery.[14]

Even before this notice appeared, word of Betty's passing had spread like wildfire through the tracking station community, as well as to Adelaide, Woomera and beyond. Although Tom could be a very hard task master at work, he was widely respected. His quirky Glaswegian humour, together with Betty's bubbly outgoing personality and the four lively kids they were raising, endeared them to just about everyone they had come across. The sense of disbelief and grief was shared by many, well beyond Tom and Betty's family circle. Among those who, even more than half a century later, could still recall how upset they were to hear of Betty's untimely death, were Tom's friend from his Woomera days, Bill Miller and his fellow student officer from HMAS *Cerberus*, Fred Lynam.[15]

After the service at Blackfriar's, Betty was laid to rest in the Catholic section of what is now Woden's lawn cemetery. As was the custom of the times, Tom did not let his children attend the burial. By the end of that week he was firmly resolved to move on, to present his stoic Glaswegian face to the world. He also set that as an example for his children. 'Keep your chin up,' his father had told him, when as a twelve-year-old, he had been evacuated from Glasgow. And that is what he now told Tommy, Marg, Nick and Danae to do too.[16]

The demands of Orroral Valley were unrelenting. After a short period off work, Tom found himself back supervising the running of one of the most complex tracking stations on the planet, as well as doing the best he could to look after his children. To some extent, the older ones were able to help the other two. But only to some extent. And he was grateful for the support of his mother, Mary, who travelled from Launceston to assist. It was a measure of the deep affection which many of the station staff had for Tom, that some of them helped out too, most notably Fritz Rehwinkel and his wife, Ella. Fritz was in charge of the station's gardens and he and Tom had become firm friends, a reflection of the latter's down-to-earth leadership style. Fritz and Ella were gentle souls who would later run Rehwinkels Garden World in the Canberra village of Hall where they would put on charity days to raise money for The Red Cross, Calvary Hospital and the Murrumbateman pre-school. After Betty's death, the Rehwinkels often minded the Reid children. It was a mark of Ella's talents in the kitchen that Tommy, Marg, Nick and Danae even looked forward to eating their peas whenever she cooked.[17]

Such haphazard domestic arrangements, however, were becoming unsustainable as 1965 drew to a close. Indeed, there were suggestions from some that the Reid children should be split up. But Tom flatly refused to entertain such an idea. At that time, one of his cousins, Agnes Barry, happened to be travelling through New Zealand. And she volunteered to postpone her return to Scotland to keep house for Tom and his kids. This arrangement turned out to be a godsend although the lack of a spare bedroom in the Hackett duplex made it an uncomfortable squeeze, reminiscent of a pre-war Glasgow tenement. Canberra was still just a big country town with less than 100 000 inhabitants, so there were few people who were unaware of the Reid family's plight. Those in charge of government accommodation made a special effort to find Tom a larger house in what was then still an extremely tight property market. EMI had arranged for the construction of two small subsidised housing estates for tracking station staff. One was in the nearby NSW town of Queanbeyan; the other in the south-western Canberra suburb of Curtin. Although rents were twenty per cent higher in Curtin, Tom settled for a largish four-bedroom brick veneer house in Macalister Crescent. Apart from providing Agnes Barry with her own bedroom and some welcome privacy, it was much closer by car to the Orroral Valley road turnoff than Hackett had been.[18]

Tom Reid's critique of Orroral's operations on that first day, had focused on the challenges of acquiring a satellite's signal. Although Tom had been able to lock onto the signal from Enos's spacecraft during its orbits over Woomera back in 1961, where its speed of 17 500 miles per hour had been about the same as those of the satellites he was now tracking, it was their sheer numbers and variety that made Orroral so challenging: up to twenty satellites per day; at different altitudes; in different orbits; and sometimes three at a time. In each case, the following procedures were adopted:

> During acquisition, the antenna, directed by data passed from Goddard [the space flight centre in Greenbelt, Maryland], searched a specified area of the sky until it picked up the beacon signal transmitted by the satellite. The antenna data system then measured and encoded the look-angles and fed this information into a servo control system and an antenna control unit in the control centre at the operations building.[19]

A further complication was the fact that no two satellites were the same. The most challenging were those in very fast low orbits such as *Pegasus*, which was tasked with measuring micro-meteorites near the Earth. Even when it moved at its top speed of 3 degrees per second, the station's 85-foot dish struggled to keep up with *Pegasus*. One of its technicians, Philip Clark, recalled watching the dish in action during a low pass.

> As the satellite came over the horizon the antenna's hydraulic drive motors could be heard increasing speed as the spacecraft approached the station. As the spacecraft reached the point of closest approach and was almost directly above the station, the hydraulic motors on the antenna were screaming like a jet engine to keep the antenna moving fast enough to keep up with the spacecraft. The whole antenna structure could be seen shaking with the movement...Then as the spacecraft approached the far horizon, the motors gradually slowed down until eventually the antenna was again stationary.[20]

On 18 November 1965, Tom hosted a visit from an American Congressional delegation, headed up by a West Virginian, Ken Hechler, the chairman of the Congressional sub-committee oversighting NASA's tracking stations. Congressman Hechler was accompanied by his colleague, Indiana Congressman Edward Roush. By this time, Orroral was successfully tracking up to thirty satellites a day, including *Explorer 25* which was gathering information on atmospheric temperature and pressure, between 500 and 2000 miles above Earth. 'It's a young man's field, and it hasn't really been tapped yet,' Congressman Hechler said. 'And the young fellows out here seem just as eager as they are back home.'[21]

While touring the operations area, Congressman Roush made use of the station's equipment to call his wife in Huntington, Indiana. The following day, *Canberra Times* reported:

> He [Roush] had only to pick up one of the special telephones at the station to reach the NASA base in Goddard, Maryland. He was then connected with his wife on a direct dialling line. The call was probably faster than most Canberra calls to subscribers on the Queanbeyan exchange.[22]

Congressional support for NASA's budget was crucial, so this visit had been important. And for Tom Reid's team it had been a triumph. This was a remarkable testament to Tom's leadership, especially given that he was still privately in deep mourning for Betty.

At First Sight

The extensive press coverage of Orroral Valley's first day's operations had aided recruitment. By early 1966, EMI Electronics had filled all the important jobs. In total about 150 people now worked at the station. This meant that it could be manned twenty-four hours a day, seven days a week, with a day shift from 8 am to 4 pm, an evening shift from 4 pm until midnight and a night shift from midnight to 8 am. Some of these shifts played havoc with family life and people complained of symptoms akin to very bad jet lag. But because orbiting satellites appeared overhead at all hours, there was no alternative. When someone called in sick, those left to cover the gap often felt put out, except when it came to the cooks. One shift supervisor, Hugh Cocking, remembered a time when the night shift cook did not turn up and the cook doing the evening shift refused to stay on. He allocated the job to Trevor 'Tubby' Devine. As Cocking recalled:

> We went through a fortnight's food in one meal. He [Devine] said there's only one thing on tonight, mixed grill, that's what you're going to get. And of course, he did everything; sausages, chops, steak, bacon and eggs, with chips. Oh boy, did the canteen manager complain the next day.[1]

At that time the station's technology was cutting edge. The computers used to decode spacecraft data, to control the 85-foot dish and to generate antenna pointing angle predictions, were the most sophisticated of their type anywhere in the world. Necessarily, the key station staff were highly intelligent and the demands on them were great. One technician Len Ahearn, was the Canberra convener of Mensa, a society whose only requirement was to have an IQ greater than ninety-eight per cent of the general population.

Even as a child Philip Clark had been fascinated by electronics and space. Prior to his appointment as a base grade station technician in 1966, he had been a senior radio communications technician with the Postmaster General's Department and then in private industry. He was experienced with radio broadcast transmitters, television transmitters, mobile two-way radios and radio communications links operating into the microwave radio spectrum. But after arriving at the station, Clark was given an introductory tour which left him astounded. 'I was amazed at the equipment and the technology,' he said. 'To me it was something like science fiction.'[2]

Tom Reid set very high standards for his team. There was no place for rote learners. He expected his engineers and technicians to be able to master complex manuals, especially on new equipment, without guidance. On Philip Clark's first day, his supervisor Arthur Meares gave him a manual, pointed to a phase-lock demodulator packed full of plug-in electronic cards and said: 'I need someone to learn how to set these up.' As Clark later recalled:

> It took me more than two weeks of continuous work to learn how to set up this complex device from the manual ... I was very much under the impression that no one at our station had actually done the full set-up and alignment procedure before ... There certainly did not seem to be anyone who could give me any guidance about it.[3]

With the station now running smoothly, the Department of Supply and NASA settled on Thursday, 24 February 1966 for its official opening or as the *Canberra Times* put it, the day upon which it was to formally go on air. For a couple of days, a major public exhibition featuring full scale models of various satellites was to be held in Canberra's Monaro Mall.[4]

In anticipation of the station's opening, a temporary raised and shaded dais for dignitaries had been built over the entrance steps to the main operations building. Tubular steel and canvas seating arrayed on the driveway in front was provided for a hundred other guests. Under each dignitary's chair was a can of Aerogard to keep the swarms of bush flies at bay. To one side was an enclosed area for the media. The most important dignitary was a seventy-five-year-old Congressman from California, George P Miller, who had flown in from San Francisco the day before and spent the night at the American embassy. An engineer by trade, Miller was the chairman of the House committee on science and aeronautics. In the mid-1950s, he had championed the creation of NASA. This jet-lagged septuagenarian was seen by the space agency as something of a father figure.[5]

While Miller had been trying to catch some sleep at the embassy, the local federal member for the Australian Capital Territory, Jim Fraser, let fly during an interview on Canberra radio. 'Traffic to and from the tracking station is causing serious deterioration to the Tharwa and Naas Road,' he thundered. 'It is not fair to the farming communities south of Canberra that these roads should be mutilated.' Claiming that the government had broken its promise to maintain them, Fraser demanded action. Whether this unsettled the minister for supply, Senator Denham Henty, is unknown. During his speech opening the station, Henty appeared unperturbed. Tom Reid managed to keep a poker face too. But privately Tom would have welcomed Fraser's criticism, having been concerned from the beginning that the station's roads' budget was woefully inadequate and that the washaways in bad weather endangered his staff.[6]

Framing the official dais were two flagpoles, one flying the Australian flag, and the other, the stars and stripes. Senator Henty's address focused on how the new station symbolised the cooperation of two countries with like ideals, while Congressman Miller stressed how NASA was charged with the peaceful exploration of space. 'It was trying to carry this out openly before the world,' he said, 'which knew of America's accomplishments but also shared in her failures.' The remaining speeches from station directors in Alaska, North Carolina and Woomera, were piped in live via NASA's worldwide communications network. Although Tom Reid was on the dais he did not speak. But later on, Tom ushered his VIP guests around, allowing Senator Henty to drive the 85-foot dish up and down at considerable speed. With the excited look of a five-year-old, the normally stern-faced sixty-two-year-old government leader of the Senate finally took his leave, full of tall tales to tell his parliamentary colleagues. This experience set a pattern and it became de rigueur for visiting VIPs to drive the dish. All the while the public had enjoyed the exhibition of full scale models in the Monaro Mall. Among these were models of six scientific satellites, including the orbiting geophysical observatory which the station was then tracking. Models of two upper-atmosphere rockets, including a Long Tom, were also on display. A day or so later, the station was opened up to the public although no one was allowed to drive the dish.[7]

Congressman Miller's comments about NASA's focus being on the peaceful and transparent exploration of space were generally true, especially in relation to manned space flight, but not necessarily so when it came to some of the satellites in Earth orbit. While an accurate picture will always be shrouded in the official secrecy laws of the United States and Australia, there are stories of strange things happening. As Alec Hempstead recalled:

> We had set up this satellite which we had never heard of before...The next thing is we were being chased out of the operations room. Then a guy came in with a briefcase chained

> to his wrist...About half an hour later, he walked out... and everything had gone, the tapes, all the chart recordings, everything all gone...There was no name given to the satellite...We were not a secret station and we didn't track secret satellites. This was a very unusual happening.[8]

After Woomera's Mini-track system was moved to Orroral Valley in late 1966, Tom's station became one of the largest and most capable tracking stations on the planet, with a capacity to support up to five satellites at the same time even though they were in different orbits. Occasionally there were up to 100 separate tracks per day. All over the world, tracking station directors wielded great power. And Tom Reid was no exception. In descending order, the station's rank hierarchy was: station director, deputy director, supervising engineer, engineer, technician, operator, clerk, storeman, and cook. Some years later each of these 'rungs of power' was parodied in the station's magazine by reference to the TV and comic book hero, *Superman*. Among them were the following:

> Station Director:
> Leaps tall buildings with a single bound,
> Is more powerful than a locomotive,
> Is faster than a bullet,
> Gives policy to God.
>
> Technician:
> Barely clears outhouses,
> Loses tug-of-war with shunting engines,
> Swims well,
> Is occasionally addressed by God.
>
> Cook:
> Lifts buildings and walks under them,
> Kicks locomotives off the tracks,
> Catches speeding bullets in his teeth and eats them,

Freezes water with a single glance,
HE IS GOD.[9]

As the tracking station progressively geared up towards full operations during 1966 and the responsibility for bringing up Tommy, Marg, Nick and Danae without Betty weighed down on him, Tom had little time for a Canberra social life. The station's location was so remote and the demands on its shift workers were so great that its employees had set up their own social club which organised children's Christmas parties and parents' nights out, as well as soccer and cricket games. Also popular were family barbecue days and autumn and winter balls. With support at home from his cousin, Agnes, Tom was able to attend some of these. But while he could be laid back and convivial, he was still seen as the director. Although he was a handsome and vigorous man who had not yet turned forty, he was also a widower with a young family of four. Only half joking he would sometimes say: 'I'm looking for a wife for my children.' Finding suitable female companionship became something of a challenge. As a station director, one of Tom's perks was to be invited to social functions at the other stations which included Honeysuckle Creek. Because it was still in the process of being fitted out, Honeysuckle's Christmas party scheduled for Friday, 16 December 1966, was to be held at the Coach House Inn Hotel in the Canberra suburb of Griffith.[10]

Woomera and the Weapons Research Institute at Salisbury had continued to play a large role in rocketry, tracking and experimentation during the mid-1960s and so there were strong links between South Australia and Canberra's rapidly growing tracking station community. Among the more recent arrivals were an Adelaide engineer, John Crowe, and his wife, Chris. Like so many others before them, the Crowes at first had nowhere to live in the national capital. So they stayed with one of their friends from Adelaide, Margaret McLachlan, who two years earlier had moved to Canberra to join the prestigious Canberra legal firm Davies, Bailey and Cater as a family law specialist.[11]

Born in 1935, Margaret had spent her early years growing up in small wheat and sheep towns north of Adelaide. Her father, Hector McLachlan, was an Elders' stock and station agent, firstly in Gladstone and then in Balaklava. After attending the local primary school, Margaret was enrolled at the Methodist Ladies College, one of Adelaide's most prestigious girls' schools. She soon developed an interest in the law and politics and was known to be a fan of Prime Minister Menzies. In 1952, the Prime Minister attended the school's fiftieth anniversary, randomly singling out Margaret for a quiet word, much to the amusement of her friends who knew how much she hero-worshipped him. By the beginning of the following year, Margaret was captain of the debating team and head prefect. That same year, aged eighteen and dressed in her school uniform, she went to the Liberal Party's local head office and joined the Young Liberals. Her parents were keen for Margaret to attend university. Her mother, Beatrice, had been thwarted in her ambition to study pharmacy by her own mother, who believed that a woman's place was in the home. Beatrice was fiercely ambitious for Margaret who wanted to study law. But Hector McLachlan did not like lawyers and was keen for his daughter to do medicine. Not wanting to openly defy him, Margaret enrolled in Arts and then quietly transferred to law the next year without telling him.[12]

Following her graduation, Margaret went to work in the Salisbury office of Scammel, Skipper and Hollidge. All the while she had continued her interest in the Liberal Party, working her way up the organisational ranks. During a luncheon break at the party's federal executive meeting in Adelaide not long before the 1961 general election, Margaret was approached by Prime Minister Menzies. 'You have to stand in Bonython,' he told her. 'There's no point,' Margaret replied. 'You'll never win Bonython and I've got other things to do.' But Menzies was insistent. 'We've got to have a candidate in every electorate to maximise the Senate vote.' As Margaret later recalled, 'Menzies needed a candidate and I was the victim.' Bonython was indeed a 'hard luck' seat for the Liberals. The incumbent, Norman Makin, an ALP stalwart who had been a cabinet minister in John Curtin's wartime government, enjoyed a massive margin. Once committed though, Margaret went at it with a will, maximising her connection with the electorate which included Salisbury where she worked, and getting the South Australian Premier, Thomas Playford, to open her campaign. In the event, Margaret gained six more votes than the Liberal candidate at the previous election. This was an excellent result given that Menzies suffered an enormous swing and came within a whisker of losing government altogether.[13]

Despite having come to the favourable attention of the Prime Minister, Margaret couldn't decide what to do with her life. Towards the end of 1964 she consulted Daniel Patrick O'Connell who was professor of jurisprudence at her alma mater. 'Why don't you leave Adelaide for a bit and expand your horizons in Canberra,' O'Connell said. 'You could do a higher degree at the ANU and because you love politics, it would give you a chance to see the federal government at close range.' Acting on this advice, Margaret took a job with Davies, Bailey and Cater at the beginning of 1965 and purchased a cosy two-bedroom house in Duffy Street, Ainslie. With the arrival of the Crowes as her house guests the following year, it was a tight squeeze and the dining room served as a combined storage area for all three of them. Attractive, vivacious and opinionated, Margaret was never short of male company, but there was no special man in her life. Her focus was on the law and politics.[14]

Towards the end of 1966, Margaret's long-term goal had begun to crystallise around gaining preselection for a winnable spot on the South Australian Senate ballot paper, her role model being Senator Nancy Buttfield, the first South Australian woman to serve in the Federal Parliament. For Margaret, this would involve a move back to Adelaide to raise her profile with potential Liberal Party preselectors. But just as she began contemplating selling her house and giving her notice to Mick Cater, Chris Crowe said, 'I've got someone for you to meet. He's a widower with four children.' Incredulous, Margaret replied: 'Chris, don't be ridiculous. I'm not interested in men with four children.' Chris backed off but she was still determined and approached Tom Reid. 'Tom, I've got someone for you to meet,' Chris said, playing cupid. 'Who?' asked Tom. 'She's a lawyer,' Chris replied. Like Margaret, Tom was incredulous: 'I need a mother for my children, not a solicitor.' And there the matter rested or so it seemed. As the day of the Honeysuckle Creek Christmas party approached, the Crowes asked Margaret to accompany them as their guest. This was their way of repaying her for putting them up as house guests, they told her, but Margaret was reluctant to go. For one thing, she had already accepted an invitation to a different function. And for another, she couldn't see what she would have in common with a bunch of engineers and technicians. But the Crowes were insistent and so she promised to drop by later in the evening.

When Margaret finally arrived at the Coach House Inn on Friday, 16 December, the Honeysuckle Christmas party was in full swing with the 'Honeysuckle Honeys', a group of trackers who held lunchtime jam sessions in the station's basement, performing Fats Waller's 'Honeysuckle Rose'. Ushered to a seat at the end of a long dining table, she found herself sitting opposite Honeysuckle's chief engineer, Wes Moon. Margaret later recalled:

> At the other end of the table was someone I'd never seen before who I noticed immediately. I started counting up the table – counting couples – and who their partners were. And I saw that he didn't have a partner. So I asked Wes who he was. 'He's the station director at Orroral Valley,' was the reply. After I pressed Wes for more information, he said, 'Marg, are you fond of children?' And when I asked why, he replied, 'Just wondered.'

Although there was no opportunity for Margaret and Tom to speak at the dinner, her first impression of him was that he was 'very attractive, manly, straight backed and in control'. The dinner was followed by an after party at someone's house.

Margaret remembers sitting on a lounge chair while Tom was perched on the arm rest talking to someone else. Then he turned to her and they began chatting. After less than a minute Tom said, 'Are you Chris's friend, the solicitor?' Looking him straight in the eye, Margaret shot back, 'Do you have four children?' A brief conversation followed. 'I don't know quite what had happened,' Margaret later said. 'But I did know that by then it was too late.' Instinctively she also knew that her plans to return to Adelaide had just gone up in smoke.[15]

John and Chris Crowe had already arranged with Margaret to invite some of the space tracking crowd over to Duffy Street, Ainslie, on Sunday, 18 December. It was to be a simple affair – drinks and nibbles from 11 am. Tom Reid was among the invited guests. And he duly arrived with his younger daughter, Danae, who was just a few weeks off turning six. Before long Tom and Margaret became immersed in deep conversation. After a while Tom asked Wes Moon to drive Danae home so that he and Margaret could have some time alone. When Danae returned, she rushed into the house and said, 'Daddy's found us a new mummy.' Startled by this, Tommy almost 14, Marg 12 and Nick 8 waited anxiously for their father. When he did finally arrive home, he stunned them with the news that he and Margaret had become engaged. Two days later, Tom and Margaret had their first date, a quiet dinner at a local restaurant. And the night after that, Margaret met Tommy, Marg and Nick over dinner at Tom's place. No one else knew. Just before Christmas 1966 Margaret flew back to Adelaide to break the news to her family. The first people she told were her mother, Beatrice, and her sister-in-law, Sandy McLachlan, who was married to Margaret's younger brother, Malcolm. Since the death of Hector McLachlan back in 1958, Malcolm, a rising young Adelaide stockbroker, had become the family's male sounding board. The first concern Beatrice and Sandy had was what Malcolm would think about a widower with four children. When Margaret told him on the afternoon of Christmas day, Malcolm's first reaction was, 'My God, how long have you known him?' To this,

Margaret parried, 'How long have I been in Canberra?' Malcolm had only one real reservation – would Tom Reid be a strong enough personality to stand up to his talented, ambitious and highly opinionated big sister? Margaret later explained her brother's concerns. 'Maybe he's someone I'll walk all over, which would not be good for me.' So early in the New Year Malcolm flew over to meet Tom in Canberra. After a tour of the Orroral Valley tracking station, the two of them went out to dinner. When Malcolm arrived back in Adelaide Tom had his seal of approval. To Margaret, Malcolm simply said, 'You won't walk over him.'[16]

And it came to pass that on 25 February 1967, Tom and Margaret were married at the Methodist National Memorial Church in Forrest, with John Crowe as best man and Sandy McLachlan as maid of honour.[17]

A Moon Made of Cheese?

All the while the construction of NASA's third Canberra tracking station in the high country above Honeysuckle Creek had proceeded apace. Approached via the Tharwa Bridge and then by the steepest road in the Australian Capital Territory, it was a difficult site to reach and the construction of its 85-foot dish presented challenges. Not the least of these were the low-loaders carrying the dish segments which had to be pushed as well as pulled up the steepest bits. The new station had just one purpose: to support manned lunar missions. As the *Canberra Times* put it: 'The road to the Moon leads through Tharwa.'[1]

NASA had thought about constructing a massive rocket to fire directly at the Moon. Instead, it chose a more fuel-efficient, multi-stage option. While drawing on the Mercury and Gemini programs, Project Apollo was far more complex. An Earth orbit which had been at the heart of those two earlier programs, was just the first of a number of steps for an Apollo mission:

1. The spacecraft would be launched atop a massive Saturn V rocket into a parking orbit around the Earth;
2. It would then be inserted into a lunar trajectory and coast for three days;
3. Next would come a braking manoeuvre to place it in orbit around the Moon;

4 After that, the lunar module would separate from the command module and descend to the lunar surface;
5 Following time on the Moon, the ascent stage of the lunar module would blast off and re-join the command module in lunar orbit;
6 Having jettisoned the lunar module, the command module would perform a burn to insert it into an Earth-bound trajectory for its return journey;
7 This vehicle would then coast for two days, with minor course corrections, before re-entering the Earth's atmosphere at 25 000 miles per hour and parachuting to a predetermined splashdown in the Pacific Ocean.[2]

Another indication of the project's complexity was the fact that an Apollo spacecraft, excluding its launching rocket, weighed over fifty tons, compared to the four-ton Gemini capsule and the two-ton Mercury capsule. As for Wernher von Braun's Saturn V rocket, it was as tall as a thirty-six-storey building and weighed 3200 tons on lift-off. Its five first stage engines, comprising the most powerful working motor the world had yet known, burned fifteen tons of liquid fuel per second and propelled it to a velocity of 5500 miles per hour. This rocket was the second most complex machine supporting America's manned missions to the Moon. The most complex was the network of stations that would be used to track them.[3]

While the existing tracking network, including a dish near Carnarvon in Western Australia, could follow the early and late phases of an Apollo mission, it was the flights to and from the Moon, together with the intricate manoeuvres on and around its surface that required three new stations to be built. These were constructed equidistantly around the globe, approximately 120° longitude apart. Their respective locations meant that as the Earth rotated on its own axis once every twenty-four hours, at least one of them would have the near side of the Moon in view, as well as any spacecraft. Only when a spacecraft was on the Moon's far side, would it be uncontactable. This network within a network comprised new purpose-built 85-foot dishes at Honeysuckle Creek near Canberra, at Goldstone near Barstow in California, and at Fresnedillas near Madrid in Spain. Each of these stations was located near a deep space tracking station, in Honeysuckle's case at Tidbinbilla. For the duration of an Apollo mission, Tidbinbilla would act as Honeysuckle's 'wing station'. When Tidbinbilla was 'slaved' to Honeysuckle in this way, it could provide full back-up and also help share the load, as when the command and lunar modules were separated and astronauts were walking on the lunar surface.[4]

Once an Apollo spacecraft had reached a point approximately 10 000 miles from Earth, whichever of Honeysuckle, Goldstone or Madrid had the astronauts 'in view', would maintain a two-way communication: an uplink to transmit commands and allow Mission Control to talk to the astronauts; and a downlink to allow the astronauts to talk back, as well as to transmit information about their heart rates, the spacecraft's fuel levels and such like. In addition, they provided a means of tracking exactly where the spacecraft was. These links that grew to almost 250 000 miles in length when an Apollo flight was in the vicinity of the Moon, were like giant electromagnetic umbilical cords upon which information travelled at almost 186 000 miles per second.[5]

As with the other tracking stations outside Canberra, Honeysuckle Creek's director, Bryan Lowe, and his deputy, Bert Forsythe, had been chosen by the Department of Supply. Both were transfers from the Weapons Research Establishment (WRE) in Salisbury. Approximately eighty others, including the chief engineer Wes Moon, were hired by a private contractor, Standard Telephones and Cables. In accordance with Australian government policy, no Americans were employed at Honeysuckle. Although many of the station's staff were British born, they were either Australian citizens or permanent residents. In essence, when Honeysuckle had the Moon in view during an Apollo mission, it would act as a communications hub, exchanging information between Mission

Control, which was almost 10 000 miles away in Houston, and the astronauts who could be up to 250 000 miles away in space or on the Moon. Except in the case of a serious glitch, all verbal communication during a Moon mission would take place between the Apollo astronauts and a Houston-based astronaut who in the tradition of Project Mercury was still known as CapCom. Accordingly, the link between Houston and Honeysuckle was just as important as the link between Honeysuckle and the Moon. To augment the Houston/Honeysuckle link, the Department of Supply took over a floor of a newly built telephone exchange in the Canberra suburb of Deakin which enabled enhanced communication via cable or Intelsat with NASA's American-based facilities.[6]

Unlike most of the senior staff at Orroral Valley who made do with training in Australia, their Honeysuckle counterparts had nearly all been to the United States in 1966 to learn of the Apollo project's progress and to meet the American-based controllers, engineers and technicians they would be dealing with during Moon missions. For the Apollo program, Honeysuckle, Goldstone and Madrid were pre-eminent among all NASA's tracking stations. Before countdown on an Apollo rocket launch could commence, each of these three stations had to be in a state of 'green light readiness'. If the status of any of them was amber or red, the engines on the Saturn V rocket could not be ignited. In a real sense these stations were as important as the rocket itself.[7]

While the various stations that tracked objects in deep space and in Earth orbit were vitally important in their different ways, none of them carried the prestige of the three Apollo stations, precisely because it was the so-called race to the Moon which captured the public's attention. Even though the Soviets had dropped off the pace a bit by the late 1960s, there was still a chance that they might attempt to cut corners by developing a new rocket to fire directly at the Moon. They would thus be able to bypass some of the intricate rendezvous and docking manoeuvres that had been built into the Apollo program. During a visit to Honeysuckle in early January 1967, one of NASA's most senior strategists,

Dr George Mueller, had warned of this. 'It is possible to make a lunar landing without a rendezvous,' Dr Mueller said. 'There are several ways of getting to the Moon.' Apart from beating the Soviets, Mueller was motivated to fulfil President Kennedys' pledge to put a man on the Moon before 1970. To make his key points to an audience of Honeysuckle engineers and technicians, Mueller used an overhead projector that required the curtains to be drawn across the station canteen's picture window. But there was a 5-inch gap between the bottom of the curtains and the floor. Just as Mueller reached his punchline, one of the station's tame kangaroos hopped by outside. Honeysuckle's operations manager, John Saxon, later recalled:

> Midway through...[Mueller's] talk I noticed a pair of kangaroo feet and hands ambling gently along from stage left to stage right...I nearly fell off my chair laughing – and so did several others who had noticed. George was not amused.[8]

Notwithstanding Dr Mueller's pep talk, it would be almost two years before NASA first attempted to send astronauts to the Moon and back. There being nothing for the three Apollo stations to track in the meantime, their respective teams were trained up using simulations. At Honeysuckle the first step was to calibrate the tracking equipment. This was done on successive nights by flying a plane back and forth over the Brindabella mountains. Displaying a white light to make it visible to Honeysuckle's technicians, this aircraft attracted a great deal of attention from concerned Canberrans. According to the *Canberra Times*, 'the control tower at the Canberra Airport...[was] inundated with telephone calls from people who reported an unidentified flying object above the city.'[9]

The tracking stations were not alone in doing simulations. On 27 January 1967, the crew of Apollo 1 was within ten minutes of a simulated launch when Mission Controllers heard one of the astronauts yell out, 'Fire. We've got a fire in the cockpit…Get us out. We're burning up…' This was followed by a shrill scream. In just twelve seconds, astronauts Gus Grissom, Edward White and Roger Chaffee were incinerated on the launch pad as molten metal dribbled down the side of their service module. Although this simulation had earlier been described as a 'very low-risk test', there had been so many glitches that Grissom had hung a lemon on the spacecraft simulator's hatch. With the Apollo program now stopped in its tracks, the command module was redesigned at a cost of $75 000 000 while the next two Apollo missions were cancelled. And Apollo 4 did not take off until November 1967. Mission Control and the wider Apollo network now doubled down on discipline and morale with a focus on two words: *tough* and *competent*. As one of NASA's leading flight directors, Gene Kranz put it:

> *Tough* means we are forever accountable for what we do or what we fail to do. We will never again compromise our responsibilities … *Competent* means we will never take anything for granted. We will never be found short in our knowledge and in our skills.[10]

Against this background, Prime Minister Harold Holt officially opened the tracking station at Honeysuckle Creek on 17 March 1967 while Vice President Hubert Humphrey was beamed in from the United States. 'This advance into space is something which the older generation like myself can scarcely comprehend,' Holt said. 'It's the stuff of science fiction made possible in our own time.' Host to the Prime Minister and to NASA's deputy director, Dr Robert Seamans, was Honeysuckle's first director, Bryan Lowe, a dapper and personable thirty-eight-year-old electrical engineer from Sheffield in South Yorkshire. In 1947, Lowe had gone to sea as a radio mechanic in a Royal Navy minesweeper after which he migrated to Australia, working at the WRE in Salisbury. On 1 April 1967, an article about Lowe appeared in Canberra's *Good Neighbour* newspaper. Claiming that the director's broad Yorkshire accent might be 'the first voice heard by American astronauts on the Moon', the paper's reporter asked him what he would say to them. 'I've got no idea what I'll say to them,' Lowe replied. 'Probably ask them if the Moon is made of cheese or something equally trite.' This article was particularly unfortunate for two reasons. Unless there was a total breakdown in communications between Mission Control and the astronauts which was only to happen rarely, as when an earthquake during Apollo 16 severed links between Houston and Canberra, no one at Honeysuckle would ever talk to the astronauts. Lowe had misconceived his role and the role of his station. And his attempt at humour would not have impressed the NASA hierarchy who were still mourning the loss of Grissom, White and Chaffee.[11]

Until NASA was ready to send a manned mission to the Moon, simulations became the focus for many key Apollo personnel. Astronauts rehearsed in spacecraft; flight controllers gamed scenarios at their consoles and tracking station technicians fine-tuned their antennae. Gene Kranz said:

> In the late 1960s our simulation technology had progressed to the point where it became virtually impossible to separate the training from the actual missions ... Simulations attempted to make events that happened in real time – malfunctions in any one of the spacecraft systems, trajectory problems, or failure of ground systems – as realistic as possible. With hundreds of possible malfunctions and many time-critical mission events, the training opportunities were limited only by the hours and weeks available to train.[12]

The tragedy of Apollo 1 had only added to the pressure. 'Nothing seemed stable,' Kranz said. 'Change was constant.' This was no less true for the tracking stations. The Americans' natural focus on checklists, on ever more intricate procedures and on fast and accurate responses, became more intense than ever before. Everything was interlinked and each step depended on every other one. The deputy director of Houston's manned spaceflight centre, George Low, described NASA's analysis of what had gone wrong with Apollo 1 as follows:

> No detail was too small to consider. We asked questions, received answers, asked more questions. We woke up in the middle of the night, remembering questions we should have asked, and jotting them down so we could ask them in the morning. If we made a mistake it was only because we weren't smart enough to ask all the right questions. Every question was answered, every failure understood, every problem solved.[13]

One consequence of this new, heightened and exacting search for excellence was that much more was expected of the tracking station directors and their key staff. Simulations were designed to ensure that every conceivable question would be asked of them and that they could provide the answers. Around Honeysuckle, NASA's visiting simulation teams were quickly nicknamed the 'What if guys'. Another consequence was the massive increase in paperwork flowing into the stations' operations buildings. Flight plans, communication protocols, maintenance manuals and computer programs were already voluminous. And many of them, in large, loose-leafed, ring-binder formats, needed to be constantly updated. To miss an update meant that any subsequent updates would not make sense. In addition, the station directors would be flooded with up to 100 telex messages per day. Some were just two lines long while others could comprise two pages. Although the longer ones tended to be more complex, a vital change to a procedure could be encapsulated in a two-line message. None of them could be ignored, discarded or delayed.[14]

It was a sure sign that all was not well at Honeysuckle when piles of paperwork began to build up on Bryan Lowe's desk. Parcels of loose-leaf inserts remained unopened, the ring binders were not updated and the telex messages remained unread.

Although the director was very intelligent, it seemed that he had an aversion to detail. As Honeysuckle's historian, Colin Mackellar put it, Lowe was 'super disorganised'. This flowed down the line to his senior staff, to his engineers and to his technicians. How could any of them be across their work when the director was not passing NASA's updates down the line to them? Only partly tongue in cheek, the Honeysuckle Honeys adapted Harry Belafonte's then popular version of the 'Banana Boat' song to express their frustrations. The final verse said it all:

> Morale is down as far as it can go,
> Five o'clock and we wanna go home,
> Station Director is also Lowe,
> Five o'clock and we wanna go home.[15]

For a short while, Lowe's charm and public relations skills were able to paper over these shortcomings. But a series of brutal simulations would soon reveal to NASA the true state of Honeysuckle's disarray. On 26 June 1967, the *Canberra Times* reported that a 'top secret' aircraft, would shortly be arriving to test Honeysuckle's readiness. This plane was a NASA-owned Super-Constellation, jam-packed with sixteen racks of equipment to generate all of an Apollo spacecraft's signals, including voice, command and telemetry. To the trackers at Honeysuckle it had the electronic footprint of a spaceship. On board were six NASA specialists. Over a week in early July, this 'Cal-plane' as it was called, made multiple passes at predetermined altitudes, backwards and forwards over the Brindabella Mountains at a height of approximately 25 000 feet. Because this plane did not fly in a well-defined trajectory, it would prove to be more difficult to track than an Apollo spacecraft. It was said that 'if you could track the Cal-plane, you could track anything'.[16]

NASA's simulation team was led by George Harris Jr, a hard-driving alpha male perfectionist who roamed around Honeysuckle's operations building with his team, firing all sorts of 'What if' scenarios to Bryan Lowe's frazzled staff. The idea was to apply maximum pressure. After all, the Honeysuckle guys were supposed to be able to pick up a twenty-watt signal from the Moon, about a fifth of the power needed to illuminate a domestic light bulb, and then through a series of complex steps, provide enough 'gain' on it to make it intelligible to flight controllers in Houston. This was difficult and delicate stuff. It soon became clear to Harris, from a combination of Cal-flights and 'What if' sessions, that the Honeysuckle team was unable to complete many of the tracking station's most basic functions, let alone cope with the high-end scenarios he had developed. And it didn't take him long to figure out that the problem was with the director.[17]

At one point, towards the end of the week-long simulation session, a clearly frustrated George Harris called everyone together in the station's downstairs canteen. As Bryan Lowe sat perched cross-legged on top of a Cottee's soft drink dispenser, Harris rose from his chair and began to speak. Born and raised in Britain, Harris was far more direct than the Americans who accompanied him. After recounting a litany of short-comings he and his team had uncovered, Harris let fly. Taking a moment to look around the room at the assemblage of engineers and technicians for maximum effect, he finally said in a strong but measured voice: 'You guys are a bunch of shit.' On hearing this, Bryan Lowe almost fell off his perch. While the ultra-polite Americans on Harris's team looked uncomfortable with his directness, they did not disagree with a word he had said. Harris, however, was not finished. After a week of hopeless yet exhausting simulations, he demanded that they continue. Upon hearing this, Lowe protested. 'You know we've been on all night,' he said, 'and we're all too tired to track any further.' From that moment Lowe's fate was sealed. The director did not seem to grasp the urgency with which NASA was pursuing operational excellence, an absolute prerequisite to achieving the goal President Kennedy had set in 1961. There was no place for a bureaucratic clock-watching attitude. What was needed was 'can do' leadership. And when Harris returned to the Goddard

Space Flight Centre in Greenbelt, Maryland, he recommended that Lowe be replaced.[18]

Pursuant to a longstanding agreement between the United States and Australia, the replacement of the director of a space tracking station on Australian soil, no matter how urgent it might be for NASA, was one for the Commonwealth government. In practical terms, this meant that the decision would be taken by MS Kirkpatrick, who was now the superintendent of the American projects division in the Department of Supply, and his soon to be replacement, Bob Leslie. George Harris Jr would also have considerable input. As they surveyed the very short list of possible candidates, Tom Reid's name stood out. Having been known to Kirkpatrick and Harris from his Woomera days, Reid had also crossed paths with Leslie when, in more recent times, they been the respective directors of Orroral Valley and Tidbinbilla. What the three of them had seen of Tom Reid, they liked: his high intelligence, his attention to detail, his capacity to think laterally, and his no-nonsense ability to lead a team of engineers and technicians. They had no doubt that of all the possible candidates, Reid was the best. As Gene Kranz would have put it, Reid was both *tough* and *competent*. On 4 August 1967, a notice was issued by the WRE advising that Bryan Lowe would return to Salisbury, that Tom Reid would take up duties as station director at Honeysuckle, and that Tom's deputy at Orroral Valley, Lou Cotrell, would take over there. In this way, Reid stepped up to a leadership role in the most advanced communications network that had yet been devised, but one that was in danger of failing because of Honeysuckle's woeful performance.[19]

A Hard Man

After transferring to Honeysuckle Creek, Tom Reid took over responsibility for a dish which looked much the same as the one he had just left at Orroral Valley. But because it tracked Moon missions rather than Earth orbiting satellites, it operated very differently. Acutely aware that his NASA bosses would want to see immediate improvement in Honeysuckle's performance, Reid realised that he had to understand a system which was new to him; a system that he believed his predecessor had never fully understood. Reading voraciously, he devoured all he could find on the Unified S-band system which was at the heart of two-way voice communication across 250 000 miles of space.[1]

In developing this system, NASA utilised that part of the ultra-high radio frequency spectrum located between 2000 and 4000 megahertz. This was ideal for space communication because it was minimally affected by the Earth's atmosphere compared to the much lower frequencies which were used by commercial television transmitter towers. The planned uplink frequency to the command module was 2106.4 megahertz. On this link, three distinct types of information could be 'unified' or carried simultaneously: the CapCom's voice from Houston to the astronauts; automatic commands to the module; and the ranging code, which by being sent up to the module and then back down again, helped to determine exactly where it was. After being transmitted by

Honeysuckle's dish and received by the command module, this uplinked information would be unbundled by an onboard computer: voice signals being converted for the astronauts to hear; commands being sent to its control systems; and the ranging code being transferred to the command module's downlink. From there on so-called sub-carriers, the ranging code would be transmitted back down to Honeysuckle, along with telemetry and the astronauts' voices for processing and on-forwarding to Houston via the Goddard Space Flight Centre. To avoid being intermingled with the command module's uplink, its downlink would use the slightly higher frequency of 2287.5 megahertz. Uplink and downlink communications with the lunar module, when it was operating independently, would work in exactly the same way except that they would be carried on slightly different frequencies to avoid being mixed up with the command module's links. With these signals travelling at almost the speed of light, there would be virtually no delay in the conversations between the astronauts and Mission Control's CapCom.[2]

Having absorbed these basics Tom set off for his first day at Honeysuckle. He was acutely aware from scuttlebutt within the Australian tracking station network and from George Harris's more formal briefings that management problems were endemic and that the station simply could not do what NASA expected of it. By way of silent confirmation, piles of papers, many of them in unopened packets, littered the director's desk. Mobilising the intense powers of concentration and attention to detail that had so impressed his professor at the University of Glasgow, Tom ploughed through them, summoning the relevant engineer or technician to explain via a series of snappy, probing questions anything he didn't understand. In doing this, he cleared the backlog within two weeks, all the while coming to an appreciation of who knew what they were talking about and who did not. Thereafter, he made a point of never having anything more on his desk than a ruler and three pencils, together with whatever he was reading, an ash tray and an occasional cup of tea.[3]

As Honeysuckle's director, Tom Reid was answerable to three masters: to the Commonwealth Department of Supply; to Mission Control in Houston; and to NASA's Goddard Space Flight Centre in Greenbelt, Maryland. It was Goddard that effectively 'owned' Apollo's worldwide space tracking network; it was Goddard that was carrying out the brutal Constellation simulation exercises; it was Goddard that on every day of an Apollo mission would test the readiness of each tracking station before signing it over to Houston; and it was through Goddard that all the data coming in from the astronauts to Mission Control via hub tracking stations such as Honeysuckle and vice versa, would be assessed for the best signals to send on. If Honeysuckle was NASA's Australian network communications hub for Apollo missions, then Goddard was NASA's worldwide communications hub, the most complex of all NASA's machines.[4]

Located on and immediately below a high ridge which sat saddle-like between two granite peaks, the Honeysuckle Creek tracking station complex was constructed on a series of terraces which had been bulldozed out of the mountain side, creating a 14-acre clearing among the eucalypts. The station's most prominent feature was its 85-foot diameter dish which had been built on the ridge. On the first terrace down was the station's headquarters, a two-storey operations building large enough to accommodate about fifty engineers, technicians and support staff per twelve-hour shift, together with the equipment that enabled them to keep astronauts on the Moon in communication with Mission Control in Houston. And on the terrace below that was a building housing a number of diesel generators which supplied all the station's power. These structures, which also included fuel tanks and carparks, were linked by manicured lawns, landscaped gardens and pathways. The complex resembled a high-tech campus.

Because the lunar module's downlink signals would be very weak, Honeysuckle's dish had been designed to use every square inch of its concave parabolic shape to 'catch' as many of them as possible. These captured signals would then be reflected up onto a small convex parabolic sub-reflector held aloft above the dish's centre by four delicate metal struts. By now more concentrated, the signals would in turn be reflected back down again, this time to receivers at the dish's centre. To achieve maximum 'gain', the receivers would be cooled to almost −270°C (−454°F). In outer space, heat equals noise. And the more such background noise could be eliminated through cooling, the greater the clarity of the signals.[5]

To generate uplink signals to Apollo spacecraft, the dish relied on two twenty-kilowatt transmitters which were located immediately behind it. When in operation, these transmitters generated significant radiation and so during tracking operations there was an exclusion zone around them. One consequence was that all the signals and commands going backwards and forwards between the dish and Honeysuckle's operations building were carried by cables in so-called 'cable trays' – trenches dug just below the surface over which metal covers were placed for ease of inspection. Despite the apparent sturdiness of these trays, rats and mice managed to find a way in to chew on the cables. But these rodents attracted snakes. And whenever Honeysuckle's maintenance team went to check the cables for rodent bites, they had to be careful not to get bitten by the snakes. It was also up to the maintenance team to maintain the hydraulic servo motors which drove the dish both up and down and from side to side – its so-called X Y axis. These motors were so delicate that the oil used in them had to be immaculate. So oil samples were regularly inspected under a microscope to ensure their purity.[6]

If the dish was the centrepiece of Tom Reid's new domain, the operations building one terrace down was its nerve centre. Just behind the main operations room window that provided a view of the dish on the ridge above was a console. And from there a technician could drive the dish by rolling a so-called 'servo ball' with the palm of the hand. This bowling ball-sized object, only part of which appeared above the desktop of the operator's console, moved through all planes like a modern-day computer mouse.

Depending on which way the operator rolled this ball with the palm of his hand, the dish would move up and down and from side to side. The speed of its movement depended on how fast the ball was rolled. While tracking though, the dish's movements mostly followed a computer program. Immediately beside this console was another, the S-band demodulator console from which among other things, raw data could be stripped off the downlink's various sub-carriers. The remainder of the operations room was divided up into areas. These included the Unified S-band operational area, the telemetry operational area, the computer operational area, and the operations console from which all these other areas were monitored and as necessary directed by the person sitting in the 'Ops 1' seat. The operations console was also the interface between the station and Mission Control in Houston. Further back were the communications switch room and the communications teletype and operations area. These operational areas were also designed to process and send on to Houston, data coming in from Tidbinbilla when it was slaved to Honeysuckle during an Apollo Mission. One of Houston's network controllers, Richard Starchurski, described part of what this system could do.

> Signals received from spacecraft are run through signal processing units that strip out the telemetry, ranging code, and voice from the composite signal. A Univac 642B computer arranges the telemetry data into pre-stored formats that can then be selected for transmission to Houston. A second Univac 642B stores digital commands that can be selectively called up by controllers in Houston for transmission to spacecraft. Another onsite processor collects tracking information including range, range rate, and the angles that show which way the antenna is pointing, and the formats for its transmission. The Unified S-band tracking system measures range with a resolution of about 5 feet, velocity to about 4 inches per second, and angles with an accuracy of 0.025 degrees. Crew voice transmissions and the processed

> telemetry and tracking data are then transmitted over long-line communications circuits to the Goddard Space Flight Centre in Greenbelt, Maryland.[7]

To facilitate such operations as a hub station, Honeysuckle was connected to Goddard by six key circuits: Net 1 was designated to carry voice communications between the CapCom in Houston and the astronauts; Net 2 was used by Mission Control's duty Network Controller to co-ordinate the network of space tracking stations; Net 3 allowed Goddard to coordinate a particular tracking station's readiness as its view period approached; Net 4 carried telemetry down from spacecraft on or near the Moon and also carried commands up to them; Net 5 carried the high-speed tracking data used to measure a spacecraft's position and predicted future trajectory; and Net 6 carried biomedical data such as the astronauts' heartbeats and respiration rates.[8]

Having done his best to understand the station's systems which for their time were so advanced that, in some respects, they were still experimental, Tom Reid set about getting to know those who worked there. Following a practice begun in the Navy, he directed one of his staff to take a mug shot of every employee. In this way word soon got around that the new director meant business. He wanted to know who they were, what they did and how well they did it. As early as 6 September 1967, Goddard's representative in Australia, John South, was able to report back to his superiors.

> I think he [Reid] is the man we needed to rectify… [Honeysuckle's] problems…All indications are that he is rapidly identifying and isolating the problem areas and then resolving them. I have discussed with Mr Reid and Bob Leslie the possibility of a senior…[Goddard] Operations representative coming over to Honeysuckle Creek for a period of one to three months to assist in the operations. Mr Reid does not desire anyone at this time.[9]

To his astonishment, Tom Reid discovered that Bryan Lowe and his deputy, Bert Forsythe, had been reluctant to sit in the Ops 1 seat. It was from this position that the station's many complex functions were orchestrated and monitored and communications with Goddard, Houston and the other stations were handled. Tom was competent to sit in this seat, even though his Glaswegian accent was challenging for Mission Control to understand and the response 'Say again, Honeysuckle' was sometimes heard on Net 2. His accent notwithstanding, the new director was determined that his deputy would be competent to sit in the Ops 1 seat too. Lowe having already departed, Forsythe's days were accordingly numbered and he was soon replaced by Mike Dinn, a British-born electrical engineer who had been deputy director at Tidbinbilla. Reid turned his attention to the chief engineer, Wes Moon, and to the other engineers, all of whom bar one or two were swiftly moved on or encouraged to resign. One of the few survivors was the maintenance engineer, Jim Kirkpatrick, who vividly recalled his fellow Scot. 'Tom was a hard man in those days', Kirkpatrick said. 'He never raised his voice. But he had thick glasses. And if he was displeased, his eyes filled the lenses.'[10]

Strictly speaking Tom had no role in hiring and firing. This was the responsibility of John Matthews, the on-site representative of Standard Telephones and Cables, a specialist communications company which had been contracted to provide all the engineering and technical staff. Matthews, however, was soon replaced by a British-born technician, Tony Cobden, who had known Tom in Woomera. As far as Cobden was concerned: 'What Tom Reid wanted, Tom Reid got.' In answer to the question: 'Who owned Honeysuckle?', Mike Dinn partly tongue-in-cheek replied: 'Tom Reid owned it because he was the Director and didn't care which Australian Government Agency might nominally be in charge.' Desperate to get Honeysuckle up to speed and to show the Americans that Australia had what it took, the Commonwealth government was happy to look the other way. In consultation with Tom and his newly appointed chief engineer, Bill Kempees, Cobden developed a questionnaire for all short-listed job applicants that apart from their technical skills, focused on initiative, motivation, problem solving and calmness in a crisis. The result was that young technicians ended up doing key jobs which in other Apollo ground facilities, were the responsibility of older fully qualified engineers, an outcome often remarked upon by visiting NASA bigwigs. Apart from one of his operations managers, Ken Lee, who was nicknamed 'the silver fox' on account of his grey hair, Tom was the oldest person at Honeysuckle where the staff had an average age of about thirty-two.[11]

No one better epitomised the result of Tom Reid's recruiting process than Kevin Gallegos, a Brisbane-born former petty officer who had learned his technical skills by operating and repairing electrical equipment aboard Australian warships. While Gallegos conceded that the Navy's systems were not as sophisticated as those at Honeysuckle and that he hadn't even finished high school, he was recruited anyway. Despite his laconic, knock-about, irreverent ways and truncated school education in outback Toowoomba, Gallegos, not yet thirty, was whip-smart, calm in a crisis and, above all, motivated, with an enormous capacity to learn and to learn quickly. Gallegos was an example of what Tom Reid had long understood: that technicians often knew their equipment better than their supervising engineers. If Tom had known that his frown of disapproval which his thick-lensed glasses magnified in an intimidating way, had been irreverently labelled 'the director's death stare' by Gallegos, this new recruit might not have prospered. But he did. Within a year, Gallegos was in charge of the all-important subcarrier data demodulator system, 'SUDS' for short, which was in effect the key switching point for all communications within the station going up to the astronauts and coming down from them. Among other things, Gallegos had to recognise all the different types of signals and then choose the best ones. Many of his colleagues agreed that Gallegos's job was the most challenging at the station. Having assumed that Tom Reid barely knew him, Gallegos was later surprised when the director asked him to accompany him on the drive back to Canberra one evening. Wondering what he'd done wrong, Gallegos was astonished when Tom began to interrogate him about the latest developments in digital voice technology. Although his approach as director was to let his managers manage and this led some of the station staff, including Gallegos, to believe that he was a remote figure, Tom knew what his technicians were up to and what they were capable of, except perhaps for the irreverent description Gallegos had given to his intimidating frown.[12]

At home, too, Tom was well capable of deploying an intimidating frown to get his way, no more so than in the early days of married life with Margaret, whom he loved dearly and respected greatly. Tom had been serious when he had said that he was looking for a mother for his children. But in Margaret's case, he acknowledged from the beginning that she had her own professional life to lead and that she would continue to work full-time as one of Canberra's leading family lawyers in one of its most prestigious firms. Margaret had the same dedication to her work as Tom did to his. 'Ever since I could remember I wanted to practise law,' Margaret said. 'There was never a time when I wanted to do anything else.' For Tommy, Marg, Nick and Danae, however, this required considerable readjustment. Having been used to Tom's cousin, Agnes, picking up after them there were new rules. Tom demanded that the kids' rooms be ship-shape by 8 am and that they share other household chores so that Margaret could get to work without running late. Fixing them with his frown Tom said, 'Margaret's got as much right to leave on time as the rest of us.' This had the desired effect even though none of them quite matched their father's ordered approach. As Danae later put it: 'His filing system for life is meticulous, his wardrobe layout ridiculous.'[13]

Within weeks of marrying Margaret and moving to a new house in Glasgow Street, Hughes, Tom flew out for a NASA briefing in the United States. Margaret soon realised that if she was not well organised she would not be able to maintain her legal career. A prompt and tidy start, with all ready to leave by 8 am, was just the beginning of a typical working day which would not see her arrive home again much before 5.30 pm. Back in the mid to late 1960s, it was more the exception than the rule for a married couple with four school age children to both be working full time in demanding professional jobs. And in those days shops did not sell takeaway or precooked meals. Margaret did it all herself, planning meals three weeks in advance, cooking them up over the weekend and then storing them, all clearly and meticulously labelled, in a large freezer. Sometimes, around 4 pm, Tom would ring his wife at work to let her know he would be bringing some visiting NASA people home for dinner. Margaret would then call home and ask one of the kids who had already finished school to defrost a particular three-course meal. By 7 pm she would be ready to serve an entrée of soup, followed by a main course of meat or fish, followed by dessert or cheese, having had young Marg's help to set the table. Other domestic chores, like washing, ironing and tidying up inside and outside the house, were handled just as efficiently, although the kids occasionally grumbled about Margaret's frozen sandwiches which they took to school for lunch every day. In addition to all the cooking and housework to be attended to on the weekend, Tom and Margaret often entertained her friends and work colleagues. During the week, Margaret also found time to attend local Liberal Party branch meetings.[14]

Although Tom Reid's early days at Honeysuckle were taken up with personnel changes and simulations, the first manned Apollo Moon mission was still some way off. Tom had more flexibility in his new job than he had been able to enjoy at Orroral Valley, where satellites needed to be tracked at all hours of the day and night. This flexibility gave him precious time to spend with his children. On Sunday afternoons, Tom would help his elder daughter with her maths homework. Marg Reid later recalled:

> Logarithms and algorithms were a blank page to me. But Dad was patient and understanding and spent considerable time setting out the numbers and explaining the process…

> [However] maths remained excruciating for me. No doubt he must have been very frustrated. His brain and mine worked on different levels.[15]

During the warmer months, Marg's younger sister, Danae, enjoyed nothing more than sitting outside on the back step at home with her dad while he studied the stars. At Tom's seventieth birthday party, Danae recalled these evenings which later developed into full blown discussions about the meaning of life:

> I would seize any opportunity I could to sit out with him during these reflective moments, because these were the most precious times, just to spend time, just talking about life and the universe, literally, just to reflect and think and learn. And all the time with that feeling, that sense, that knowledge, that life like the universe is infinite, that there are no boundaries, and there is still so much to discover...principles not unlike Dad's broader philosophies on life.[16]

Santa Claus

Just as Tom Reid was assessing the team at Honeysuckle, they were assessing him. One of the engineers, John Crowe, remembered Reid's cut-through style:

> Tom had confidence in his own ability. He didn't suffer fools and had a steel trap of a mind. He would assess his options very quickly and then say, 'this is what we are going to do'. No one challenged him. He was on top of it. And there was nothing on his desk when he went home because he liked to clear up problems before he left for the evening.[1]

The station's union representative, Bryan Sullivan, was a Sydney-born computer software specialist, a rare breed in those days. Although he didn't always agree with the new director, he liked and respected him. About a week after Sullivan's mug shot had been taken, Tom Reid suddenly appeared beside his console and without any introductory small talk, proceeded to quiz him about his job experience, which included working up the electrical systems of refurbished warships at Sydney's Cockatoo Island dockyard. Sullivan appreciated Reid's management style, describing him as a man of few words who wanted clear and precise answers:

> One day Tom asked me about a complex computer problem. And when I began to start into the detail, he said: 'I don't need to know all that digital waffle. When will it be fixed?' I quickly realised the difference between senior management and junior engineers.[2]

Reid often used humour to make his point. To combat waffle, he sometimes referred to a technical writing kit known as the 'Buzz Phrase, Buzz Word Generator'. This was made up of multiple lists of ten phrases such as: 'on the other hand', 'the fully integrated test system', 'is further compounded when taking into account', 'the total system rationale', and multiple lists of ten words including: 'responsive', 'transitional', 'organisational' and 'flexibility'. And it came with an explanation:

> This kit is based on the Simplified Integrated Modular Prose (SIMP) writing system. Anyone who can count to 10, can write up to 40 000 discrete, well-balanced grammatically correct sentences using the buzz phrase section or the same number of incomprehensible intelligent sounding technical terms with the buzz word generator.[3]

Known for his 'crisp management style', Tom Reid saw jargon as the enemy of clear thinking and effective action especially in a crisis. For the Apollo program he believed that the only safe course was to keep in mind Murphy's Law: 'Anything that can go wrong, will go wrong'. This was also the philosophy of George Harris's simulation team, which was tasked with testing Tom the same way it had tested Bryan Lowe, by 'messing up the smooth flow of pass procedures' as one technician, Hamish Lindsay, put it. To give Lindsay and his other young technicians like Kevin Gallegos and Bryan Sullivan an edge, Tom encouraged his deputy, Mike Dinn, to develop a simulation capability within the station, by using any surplus equipment he could lay his hands on to replicate the signals that would be sent and received during a real

Moon mission. In July 1968, Dinn travelled to Fremantle where an obsolescent NASA tracking ship, the *Coastal Sentry Quebec*, had just been sold for scrap. While many of the parts Dinn salvaged, including complete operations consoles, were later used at the station to fit out a unique simulation room, his immediate priorities were a number of meters which were installed on Honeysuckle's main operations console. These were used to measure the command and lunar modules' signal strengths as they came out of Honeysuckle's receiver. The other Apollo stations at Goldstone and Madrid had been fitted out by the Bendix Corporation which had an aversion to adding or altering any standard NASA equipment, let alone building a simulation room. Not so with Tom Reid who as he grew in confidence would turn a Nelsonian blind eye to NASA procedures when it suited him.[4]

This created in some of Reid's more innovative team members an appetite to do the same. A case in point was Bryan Sullivan who wrote all the software for simulation training. To achieve the greatest authenticity, Sullivan needed to be able to replicate NASA's most sensitive and secure line, the incoming command line from Mission Control which was highly protected against the possibility of a Soviet hack. Without ever telling his director, Sullivan managed to break into this line to access a complete picture of NASA's communications. This, in turn, allowed him to write software replicating any command Honeysuckle needed for its internal simulations and made information coming from the next room look like it was coming in from Houston. 'I couldn't let on to Tom,' Sullivan later said, 'because he would have sacked me'. More likely, Tom would have turned a blind eye but only because Sullivan had succeeded without being detected.[5]

Some simulations tested the human element. Hamish Lindsay recalled being timed by a stopwatch as he was sent running to the station's store for a spare part. During one test, Tom's operations manager, John Saxon, pulled a module out of a microwave link while the operator was on a brief toilet break so as to simulate the failure of a key component. And when the operator returned and eventually figured out what Saxon had done, he was so incensed that he backed Saxon up against the nearest wall. As part of another simulation, a Honeysuckle technician feigned a heart attack so convincingly that his offsider raced off outside to vomit. Once mastered, however, such challenges prepared Tom's team for just about any eventuality.[6]

In November 1967, Honeysuckle was involved in tracking Apollo 4, the first test flight of a Saturn V rocket. This was an unmanned two-orbit mission, the success of which after the disaster of Apollo 1, raised hopes that NASA might yet be able to land a man on the Moon before 1970. Honeysuckle's role was limited to tracking the first orbit. The Apollo 5 and 6 missions which took place in January and April 1968, were also unmanned orbital flights. As such, they did not test Honeysuckle's ability to track spacecraft to the Moon and back. But they did allow the wider network comprising the three Apollo tracking stations, Goddard and Mission Control to work through some real-time flights together. And Honeysuckle, which was the only Apollo station not to have American personnel embedded in its team, performed well.[7]

Honeysuckle also performed well for George Harris's Cal-plane simulation team. Importantly, Tom Reid had the strength of character to stand up to Harris who had run rings around the unfortunate Bryan Lowe. It helped that Reid and Harris had known each other since Woomera days, that they shared a British sense of humour and that they enjoyed a Scotch or two. A sense of humour was especially important when some Honeysuckle staff decided to play a trick on one of Harris's team, Robert 'Robby' Burns, who later recalled:

> I was listening to the various in-house voice channels when I heard a sugary-sweet young lady's voice in my headset saying, 'Robby darling … oh Robby daarling …' I broke into a cold sweat as the voice kept getting more personal by the second. I kept thinking that everyone could hear what was going on and I couldn't figure out how to get her to stop. I just wanted to

> crawl behind the cabinets and hide...The operations people later assured me that I was the only one who could hear her.

Such humour was not a sign of weakness at Honeysuckle; it was a sign of strength. After one particularly testing simulation, Tom Reid, accompanied by the station's kangaroo mascot, presented George Harris with a pair of sandals mounted on a wooden plaque. The inscription below read: 'To Mr G Harris Jr in the hope that his feet will be dry when walking on water.'[8]

None of the other tracking stations developed anything like Honeysuckle's simulations. And in July 1968, George Harris was able to report back to NASA that Honeysuckle was 'the best station in the network'. In less than twelve months, it had moved from the bottom of the pile to the top. Tom had a well-deserved reputation for not letting NASA push him around. If he didn't like a procedure NASA was trying to impose, he'd let Goddard know in typically blunt fashion. As often as not he'd do things his own way. During Apollo missions the tracking stations were each to be manned by two twelve-hour shifts. At the American-run stations, there was to be a prime shift when the Moon was in view and a back-up shift when no signal could be received. But this led to an 'A' team and 'B' team mentality. At Honeysuckle Tom did things differently, insisting that what were called the operations and maintenance shifts were each able to work as well as each other, to tackle any problem that might arise. During simulations both shifts were expected to operate flawlessly. This meant that in a crisis either Honeysuckle shift could take the initiative; unlike Goldstone and Madrid therefore, neither of Tom's shifts had to wait for instructions from Houston. Bryan Sullivan described what it meant for those on the front line.

> As technical staff, each of us had to know as much as possible about every aspect of space tracking operations. In United States tracking stations there were enough staff for every member to be able to specialise. But in Australia we all had to

> be familiar with as much equipment as possible and be able to do other jobs in an emergency. Our roles were much more multi-skilled...The Americans worked intensively on a very special task. We were expected to understand the broader picture.[9]

The American Central Intelligence Agency, meanwhile, had reported that the Soviets were working towards a manned lunar flight. Using a powerful new Soyuz Zond spacecraft, two cosmonauts would circle the Moon without going into a lunar orbit or attempting a Moon landing. George Mueller's prediction made in January 1967, that the Russians might cut corners and fire a powerful rocket directly at the Moon, appeared prescient. This triggered frenetic debate within NASA about whether the Soviets could be trumped with a manned orbital flight, not just a circumlunar loop around the Moon's far side. 'We don't even know if we can compute a lunar injection manoeuvre,' Gene Kranz said. 'Christ, we don't even know if the booster guidance can do the job.' Nevertheless, the decision to go to the Moon on the next mission but one was duly made in August 1968, even before the first manned flight of an Apollo spacecraft had taken place. A few weeks later, the Soviets launched another Zond spacecraft which successfully completed a loop flight around the Moon. Although its payload of turtles, mealworms and wine flies survived their very steep re-entry deceleration rate, it would probably have been lethal for cosmonauts. While the Soviets were again ahead in the race for the Moon, they still had many problems to iron out.[10]

The first three men to ride into space atop a massive Saturn rocket were led by the veteran astronaut, Wally Schirra. Over almost eleven days in mid-October 1968, they orbited the Earth 163 times. All of NASA's objectives were met, including some successful tracking by Honeysuckle and its wing station at Tidbinbilla. The stage was set for a manned flight to the Moon including a lunar orbit. Most noteworthy was an innovation that in a technical sense was superfluous to the Apollo program – live television.

The first astronaut to film with a TV camera in space, although not for a live broadcast, had been Gordon Cooper during his Mercury flight in early 1963. 'I think most of us felt it was important to personalise the flight, that's what really made the public get close behind us,' Cooper said following the broadcast which went to air after his return. But for NASA's managers, TV had only been necessary to monitor the astronaut and the spacecraft.[11]

By the middle of 1968, NASA's technology had advanced to the point where a live TV broadcast from Earth orbit was possible. Under intense pressure from America's commercial television networks, NASA agreed to attempt such a broadcast during Apollo 7. But Wally Schirra remained sceptical. 'We're going to be too busy,' he said, 'to provide the world with a vicarious thrill.' And after achieving Earth orbit, he defied not only the CapCom on duty but also his ultimate boss, Deke Slayton, NASA's senior astronaut who was known as CAPCOM Number 1. When Houston contacted Schirra with new procedures to power up Apollo 7's TV camera, he mutinied:

> SCHIRRA: 'I can tell you at this point TV will be delayed without any further discussion until after the rendezvous.'
>
> CAPCOM: 'Roger. Copy.'
>
> SCHIRRA: 'Roger.'
>
> CAPCOM: 'Apollo 7 – This is CAPCOM Number 1.'
>
> SCHIRRA: 'Roger.'...
>
> CAPCOM Number 1: 'All we've agreed to do on this particular pass is to flip the switch on. No other activity associated with TV; I think we are still obligated to do that.'

> SCHIRRA: 'We do not have the equipment out; we have not had an opportunity to follow setting; we have not eaten at this point. At this point, I have a cold. I refuse to foul up our lines in this way.'[12]

Later, Schirra had a change of heart. And for the short periods when his Earth orbiting spacecraft was in view of specially equipped tracking stations in Texas and Florida, Apollo 7 broadcast live. Among other things, Schirra held up a cue card. 'Deke Slayton,' it read, 'are you a turtle?' For this and their other spirited performances, Schirra and his crew were presented with an Emmy Award. But none of them ever flew in Space again.[13]

Having played a pivotal role as a flight director during Apollo 7, Glynn Lunney refused to confirm press speculation about whether the next mission, scheduled for December, would carry men to the Moon. But the decision to task the Apollo 8 crew with a lunar orbital flight had already been taken. According to the director of flight operations, Chris Kraft, 'it was the boldest decision we made in the whole space program – period!' A number of planned flights were now rolled into one. As the flight dynamics officer, Jerry Bostick, put it:

> From a trajectory view-point we had to accelerate some of the software ... in Mission Control, in the spacecraft, and in the worldwide tracking network. Now management had decided to go into lunar orbit, it required very accurate calculations. Shooting for the Moon is a bit like duck hunting – you don't shoot at the duck, you shoot at a spot in front of it and let it fly into it. So, you have to aim at a spot in front of the Moon equivalent to the thickness of a sheet of paper when viewed from Earth. We were a little nervous about doing it for the first time and much earlier than planned.[14]

The margin was so tight that an error of just one mile per hour in the spacecraft's speed would mean missing the Moon by 1000 miles. According to a NASA engineer: 'It was like running across the front of a speeding locomotive, close enough to get a paint sample off its front without getting hurt.' The Soviets had planned to launch a manned Moon mission before the Americans. But on 19 December, two days before the Apollo 8 launch, they were forced to postpone due to a number of technical problems. And the cosmonauts were left to watch the Saturn V launch on television. The Soviets were not the only ones to have faced technical problems. In the lead up to Apollo 8, Honeysuckle Creek had been besieged by moths attracted to the station's lights. These insects hid out in plenums through which high speed fans circulated cooling air to sensitive machinery. Neil Sandford later recalled:

> We experienced some embarrassing intermittent power amplifier trips which took some tracking down...The problem was traced to large Bogong moths that found shelter in the plenums. They would hang on in the high airflow for as long as possible, eventually let go, trip the airflow switch in passing and then be splattered by the fan...There were some interesting questions from Mission Control about fault reports which read: 'Pa trips caused by Bogong moths.'[15]

The solution was to keep lighting to a minimum in the power supply room and to fit gauze screens to all plenum areas. At the beginning of December 1968, Tom Reid received a most gratifying letter from Bob Leslie, the superintendent of the Supply Department's American projects division.

> I would like to express my congratulations to you and your staff for the accomplished manner in which Honeysuckle Creek supported the Apollo 7 mission. Considering the poor state of readiness of the station when you took over about a year ago, the difficulties inherent in working with a changing network such as the Manned Space Flight Network, and

> the rather severe contractor problems experienced at the station, the performance during Apollo 7 was remarkable ... Although there is still much to be done to ensure equal performance on the lunar phase of the Apollo mission, I have every confidence in your ability to achieve success.[16]

Bob Leslie's praise was timely because NASA's decision to turn Apollo 8 into a lunar orbital mission placed enormous pressure on Tom Reid and his team. Stepping up to gruelling twelve-hour shifts, they laboured to ensure that Honeysuckle and its wing station, Tidbinbilla, would fit seamlessly into the worldwide tracking network. For Reid, this meant digesting mountains of telexes, manuals' updates, flight plans and calculations that now flooded into the Station, before farming them out to his engineers and technicians, all the while ensuring that this paperwork was being understood and acted on. In addition, Reid was having regular conference calls with his fellow station directors at Goldstone and Madrid, with Goddard and with Mission Control. Most importantly, he worked with Tidbinbilla's director, Don Gray, to ensure that their two stations operated as one. Mike Dinn later recalled: 'Don was an unflappable space tracker who never let problems, regardless of their severity, spoil his day. He was one of the best.' Reid was lucky to have Gray as his wing man. A dress rehearsal of the Apollo 8 launch was held on 14 December, beginning in the early afternoon when Honeysuckle's equipment underwent a series of tests. At 5.30 pm, communications were established with the whole manned space flight network, followed by a countdown and simulated launch of the Saturn V rocket. All went well and the real launch took place just before 8 am on 21 December, Florida time. A day into the Mission, the *Canberra Times* reported:

> Tidbinbilla and Honeysuckle Creek...are working as one complex, tracking for 12-hour stretches...They will begin today's phase at 10.45 am. Honeysuckle receives continuous radio signals and data which is processed by

> computers and sent back to Houston through the Overseas Telecommunications Centre, Deakin, and the NASA communications network.[17]

Tom Reid elected to stay over at Honeysuckle for the duration of the mission, which lasted a week. Unlike earlier Apollo missions when Earth-orbiting spacecraft were in view of Honeysuckle for only a few minutes per orbit, Apollo 8 was in view for up to twelve hours at a time. For much of that Reid sat in the Ops 1 seat. At other times, especially in the early hours of the morning, he would suddenly appear with a cigarette in hand to watch over those working the operations console. With his face covered in stubble and wearing only shorty orange pyjamas, he looked ferocious as he peered through horn-rimmed glasses that magnified his eyes. 'No one showed any sign of amusement,' Bryan Sullivan said. 'It would have been more than their job was worth.' At one point, Honeysuckle's antenna feed cone began to play up. And while it was being replaced using a large crane and cherry pickers, Tidbinbilla seamlessly took over the tracking without breaking the link.[18]

For Apollo 8's astronauts, Frank Borman, James Lovell and William Anders, the mission was a series of firsts: the first humans to travel to the Moon and back; the first to orbit the Moon; the first to see its far side, which according to Borman 'looked like the burned-out ashes of a barbecue'; and the first to see the Earth suspended in space. No one had ever travelled further or faster or been so isolated from the rest of humanity. And it was Honeysuckle which provided their link to Mission Control at critical moments: when they first disappeared behind the Moon; when they reappeared; when they executed a rocket burn for their return to Earth; and when they re-entered the Earth's atmosphere at 25 000 miles per hour.[19]

Like Wally Schirra before him, Frank Borman's objections to taking a TV camera on Apollo 8 were overruled. With TV signals now capable of being processed through Goldstone and Madrid, although not Honeysuckle, much longer live broadcasts became possible including one of the Earth suspended in sea of blackness. The astronauts also put on a bit of a comedy show, although the timing upset many football fans. As Borman related it:

> We all hammed it up a bit. But after we got back, I heard that when CBS interrupted the pro football playoff game between the Vikings and Colts for our brief broadcast Sunday afternoon, the network had been swamped with protesting calls. Maybe we should have thrown a football around.[20]

One of the most critical steps in the whole mission was the lunar exit phase, during which Apollo 8's rocket would be fired to thrust it out of lunar orbit. This manoeuvre was to be conducted behind the Moon when Apollo 8 was out of sight and out of contact. After a rocket burn lasting a little over three minutes, the spacecraft would accelerate from 3000 to 6000 miles per hour. But any misfiring would condemn the astronauts to orbiting the Moon until their oxygen ran out and they died from asphyxiation. On Boxing Day 1968, the *Canberra Times* reported what happened:

> After the firing, there was a delay of about 10 minutes before the capsule came 'into view' and operations controllers and assistants at Honeysuckle Creek and Tidbinbilla received the astronauts' message: 'Please be advised that there is a Santa Claus.'[21]

The Mop Handle

After voyaging almost 250 000 miles to the Moon, Apollo 8 had been less than half a mile off target when it finally arrived there. And after a total round trip that more than doubled the distance when its lunar orbits were included, the spacecraft had splashed down uncomfortably close to its aircraft carrier recovery team, which is to say right on target. As a result, NASA now felt more confident that President Kennedy's deadline could be met. But there were still two missions to go before a lunar landing could be attempted: Apollo 9, which would flight test a lunar module in Earth orbit; and Apollo 10, which would be a full-dress rehearsal for Apollo 11, apart from the lunar landing. Necessarily, the scheduling of these missions was brutally tight. The eleven-day Apollo 7 mission had taken place in mid-October 1968 and the week-long Apollo 8 mission during late December. Now Apollo 9 was scheduled for ten days in early March 1969, Apollo 10 for just over a week in late May and Apollo 11, also for just over a week, in mid-July. At Honeysuckle Creek Tom Reid and his team were juggling with what, from a Mission Control perspective, Gene Kranz described as follows:

> Mission planning and preparation takes about a year with the final training starting about three months before launch. The objectives of each mission were vastly different from the

> preceding mission and now with launches spaced at two-month intervals ... everyone was working several missions simultaneously, constantly juggling schedules and priorities. The workload was punishing. Sixty to seventy-hour work weeks became commonplace.[1]

In the months following Apollo 7, Tom Reid had seen very little of his family. There was no time for maths coaching and evenings on the back step at home stargazing with Danae were just a pleasant memory. During the quieter months before preparations for Apollo 7 had begun in earnest, the children had visited Honeysuckle with their father and had some sense of the importance of his work. Ten-year-old Nick also kept a scrapbook, carefully cutting and pasting articles from the *Canberra Times*, which published numerous news stories about the local tracking stations, as well as some longer pieces by Bruce Juddery. Having first taught himself, Juddery explained the complexities of space tracking with a fluid writing style that gave Canberrans easy-to-read insights into the crucial role Honeysuckle was playing in the Apollo project.[2]

Like everyone else involved with Apollo 8, Tom Reid's 1968 Christmas had passed in a blur of frenetic activity at Honeysuckle. The only concession to the festive season had been a Christmas dinner with all the trimmings laid on by the station's catering staff. This feast had been served all day to allow the engineers and technicians to enjoy it, regardless of their respective shifts. At one point, Tom's wife Margaret sent him a Christmas card to remind him that she was still safe back home. By way of compensation the American ambassador to Australia, William Crook, offered to host a late Christmas party for the trackers and their partners at his official residence. Billed as a 'splash down party' the initial guest list was limited to sixty. But the partners of those trackers who hadn't been invited flooded the embassy's switchboard with complaints. And the number of people who eventually attended, on 8 February 1969, numbered in the hundreds. It was the largest party ever held at the American embassy, complete with marquees, the Duntroon band and generous servings of alcohol. One of the Honeysuckle engineers, John Saxon, recalled asking a massive US Marine for a gin and tonic.

> 'Yuss Sir', the Marine in full dress uniform replied, and produced a tall glass somewhere between a midi and schooner size, put in a few ice cubes and filled the glass to within an inch of the top with Tanqueray gin, followed by a splash of tonic.[3]

As the Apollo program moved into high gear for the first lunar landing, this party provided a rare opportunity for Margaret and Tom to socialise together. In an extensive interview published by the *Canberra Times* on 7 January 1969, Margaret was described as a 'space widow'. Even when her husband found the time to stay at home for a night or two, tracking station business intruded. Explaining to the reporter what this was like, Margaret said:

> How would you like to be disturbed at all hours of the day and night with overseas telephone calls…especially between 2 am and 6 am, which must be when the people in the States get into action?… There is a close link with the American stations on many jobs, but they don't seem to realise what time it is over here when they call.

Acknowledging that whatever strain she felt through Tom's absences was nothing compared to what the astronauts' wives had to go through, Margaret nevertheless worried about her husband's long drives home on unlit, potholed country roads especially after twenty hours straight at the station. But it was comforting to know that the station's food was nourishing enough for Tom to avoid feeling run down. And Margaret was pleasantly surprised at the amount of scientific knowledge, including some computer language, that she was able to pick up from Tom and his friends.[4]

For Honeysuckle Creek and Tidbinbilla, Apollo 9 presented

an unusual challenge. Having been set up respectively to track Moon and deep space missions, neither station was ideally rigged to track objects orbiting the Earth. Yet during this mission, they would be required to establish and maintain links with the command and lunar modules after they had separated during Earth orbit. The problem was that Honeysuckle couldn't track both of them at once when they would be up to 100 miles apart; and Tidbinbilla's dish wasn't fast enough to track either of them. For this particular mission only, Tom Reid and Don Gray would have to find a way, somehow or other, to make the stations work in unison before reconfiguring them for Apollo 10. It demanded just the sort of juggling that Gene Kranz faced back at Mission Control. In the end, Reid and Gray settled on an unorthodox plan to add Honeysuckle's acquisition antenna to the array. This combination would allow the lunar module to be tracked just long enough as Gray put it, 'for the required commands to be blasted in'. It was a measure of NASA's faith in the competence of Reid and Gray that their unorthodox proposal was accepted by Mission Control. As Bryan Sullivan said:

> The teams at Honeysuckle Creek and Tidbinbilla were not frightened of finding an unconventional solution. And we did manage it, with some very clever procedures. The antennae of both stations did things they were never designed to do.[5]

Weighing over 3250 tons, Apollo 9 blasted off on 3 March 1969, its Saturn V rocket having gulped down twenty-three tons of kerosene and oxygen even before it had begun to move an inch. Out of a total of 151 Earth orbits, Honeysuckle and Tidbinbilla supported fifty-two. These included four crucially important passes of the command and lunar modules in separate flight, beginning on the evening of 7 March. For well over 6 hours, astronauts James McDivitt and Russell Schweikert tested out the lunar module knowing that if anything prevented them from redocking with the command module, they and their little bug like spacecraft would burn up during re-entry. As it turned out, Honeysuckle and Tidbinbilla were able to track each module in separate flight while the overall mission which ended on 13 March was judged a great success. Now the stage was set for Apollo 10's dress rehearsal of everything except a lunar landing. Just over a week before Apollo 9's flight, the Soviets had attempted to launch an even bigger rocket, the N-1, powered by thirty engines and designed to fly cosmonauts directly to the Moon. But a minute after lift-off during an unmanned test flight, two of its engines malfunctioned and its controllers had to destroy the rocket remotely. Disintegrating into a fiery inferno with as much force as a nuclear weapon, it sprayed flaming bits and pieces over a distance of 35 miles. Any chance the Soviets had of beating the Americans with a manned lunar landing of their own before 1970 went up in smoke.[6]

Over the next couple of weeks, Tom Reid and Don Gray raced to reconfigure their stations, this time to track the separate flights of the command and lunar modules in orbit around the Moon. Everything had to be ready at least three weeks before the launch of Apollo 10 on 18 May, at which time the worldwide tracking network would be placed on 'mission status' meaning that nothing could be changed without NASA's express authorisation. In late April, Howard Kyle, one of the key designers of the Unified S-band system that linked the tracking stations with the Apollo astronauts, paid Honeysuckle a visit. Kyle later recalled:

> I was tremendously impressed by the quality of the technical people ... and their eagerness. Tom Reid and his two deputies take turns manning the operations consoles ... These people are very conscientious and intense in their efforts ... Monday 28 April: I went to Honeysuckle with Tom Reid. He collected his communications team and we spent until about 3 pm discussing problems ... It is very stimulating dealing with these people. Then we made a mad dash to Canberra airport via the Tidbinbilla Fauna and Animal Reserve. Never did see

> a kangaroo, or a koala bear, just one little wombat. Not worth the wild ride.[7]

One week out from launch, the staff at Honeysuckle and Tidbinbilla were stepped up to twelve-hour shifts so that the stations could be manned around the clock. In the days remaining, all the systems at Mission Control and at the key Apollo tracking stations were put through a series of rigorous simulations and cross-checks. A day or so before the launch, Tom Reid's repeated warnings to his team about Murphy's Law were illustrated in an incident that Bryan Sullivan described as follows:

> One evening during testing the cleaner accidentally bumped the mop handle against a 'button' on a rack of electronic equipment in the telemetry area. We all scrambled to investigate the sudden loss of data while Mission Control demanded to know what had happened.[8]

During its flight which lasted from 18 to 26 May, Apollo 10 executed thirty-one lunar orbits. Its lunar module, *Snoopy*, separated from its command module, *Charlie Brown*, and descended to within 10 miles of the Moon's surface. On board *Snoopy*, were Tom Stafford and Gene Cernan while John Young waited alone in *Charlie Brown*. As *Snoopy* passed over the proposed Apollo 11 landing site at 6000 miles per hour, Stafford took photographs. Then just as they were making their second pass, *Snoopy* began to tumble uncontrollably, such that Cernan could see the lunar horizon looming up into his window 'about five times, from different directions, in about eight seconds'. After jettisoning *Snoopy's* descent stage prematurely, Stafford managed to re-establish control. And after what remained of *Snoopy* settled back down into controlled flight, the astronauts realised that a switch in the navigation system which had been set incorrectly, was the cause of the near disaster. It was another case of Murphy's law in action. But there was nothing intrinsically wrong with Apollo 10's systems.

And the tracking stations including Honeysuckle and Tidbinbilla, performed flawlessly. The way was now clear for Apollo 11 to land astronauts on the Moon.[9]

All the while, the Soviets kept trying to steal just a bit of the limelight from what they knew was imminent: a manned lunar landing by the Americans. But their two final attempts, another N-1 rocket launch and a lunar probe which was designed to robotically drill out a core sample of Moon rock and return it to Earth, ended badly. In an explosion that was measured 2500 miles away in Sweden, the N-1 disintegrated before lifting off, obliterating its launching complex in a liquid fireball, while the lunar probe crashed into the Moon's surface and was blown to smithereens before it could begin drilling. In the early days, beginning with *Sputnik* in 1957 and then, most famously, with Yuri Gagarin's orbital flight in 1961, the Soviets had forged ahead. This was partly because the Americans insisted that space travel be made as safe as possible. NASA's first orbital flight by John Glenn the following year had received far more back-up than Gagarin's had. In short, the Soviets cut corners which saved them time. During the relatively simple early missions, fast-paced Soviet improvisation led the way. But as the complexity of putting a man on the Moon and returning him safely to Earth came to be better understood, the Americans' more methodical approach gave them the advantage. Testament to that was NASA's steady progress after Apollo 1, compared to the Soviets' ignominious efforts to catch up especially with their N-1 rockets.[10]

An essential element of the Americans' planning was the principle of redundancy, an engineering concept designed to incorporate multiple back-up systems wherever possible. Today the word redundancy is better known for its industrial meaning, related to being laid off. In its application to the Apollo missions, redundancy meant that Apollo 11 had not always been preordained to be the mission which would land men on the Moon, any more than it had been preordained that Neil Armstrong and Buzz Aldrin would be the first men to walk on its surface. This could have been attempted by Tom Stafford and Gene Cernan during Apollo 10, or delayed until Apollo 12, in which case the honours would have gone to Charles Conrad and Alan Bean; or any combination of them might have been chosen for Apollo 11. In short, there were always multiple back-ups which gave NASA options. It was the same with the tracking stations. Depending on the timing of a mission, a change in the flight plan or indeed a technical glitch, any one of the three tracking stations could become the main or 'prime' communication link with astronauts on the Moon. While there was always plenty of advance planning, there was also a touch of serendipity in the way a Mission might play out.

As between Honeysuckle and Tidbinbilla, the tracking arrangements for Apollo 11's command and lunar modules were to be shared in the same way they had been for Apollo 10. When the spacecraft were both in view and being flown independently, Honeysuckle would track the command module while Tidbinbilla looked after the lunar module. Because Tidbinbilla normally tracked objects way out in deep space, it had a special Maser receiver which was slightly more sensitive than the one attached to Honeysuckle's dish. This meant that the quality of the signal Tidbinbilla could pass on was slightly better. And because the command module's four-dish high-gain antenna transmitted a much more powerful signal than the lunar module's small single dish antenna, the more sensitive of the two tracking dishes was assigned to link up with the weaker signal.[11]

Apollo 10 had been challenging enough. But the plans for Apollo 11 required astronauts to land on the Moon, to walk on its surface and to take off from there. In terms of complexity, Apollo 11's lunar phase was a quantum leap. Anticipating this, NASA had been looking to build extra redundancy into Canberra's Apollo tracking network. It had sounded out John Bolton, the director of the CSIRO's radio telescope at Parkes in north-western New South Wales. Although Parkes had no transmitter and could not send signals up to the spacecraft, the size of its 210-foot diameter dish would enable it to receive more of the lunar module's weak downlink signal, enabling it to enhance the quality of the data passed on to Mission Control. In this way, Parkes could provide a downlink back-up to the Canberra tracking stations. And in late 1968, a one-line contract had been signed off. It read: 'The Radio Physics Division would agree to support the *Apollo 11* mission.'[12]

While Parkes was onboard in principle, the work of fitting out its dish with the necessary NASA communications equipment and of connecting it to the Australian tracking network was delayed by everyone else's intense focus on Apollo 7, 8, 9 and 10. And it was not until June 1969 that Parkes was linked up to Honeysuckle via Mount Canobolas, Sydney and Williamsdale among other places, utilising a series of microwave and cable connections. Parkes's incorporation into the network had been so late that there was no one to spare from Honeysuckle or Tidbinbilla. NASA therefore flew out some of its own people from the States to embed at Parkes for the duration of the Apollo 11 mission. In this way the Australian tracking stations for Apollo 11's all-important lunar phase now comprised Honeysuckle, Tidbinbilla and Parkes, with Tidbinbilla being connected to Honeysuckle by a microwave link. But out of those three, it was Honeysuckle which was the so-called hub station. The downlink signals gathered in by the other stations would all be routed through Honeysuckle's complex systems, enabling Tom's team to strip off the different messages before processing them, selecting the best ones and then packaging them up in a form suitable for transmission via undersea cables to Goddard and thence to Mission Control in Houston. In Tidbinbilla's case, the uplink commands, voice and tracking data would come through Honeysuckle too. Although Tom Reid, Don Gray and John Bolton were all highly accomplished professionals, it was Reid who, as director of the hub station, was first among equals. And because of his seniority and experience in space tracking, it was Reid who the Houston-based mission controllers would want to talk to if something went wrong at the Australian end during lunar tracking, whether it be at Honeysuckle, Tidbinbilla or Parkes.[13]

The tracking stations' key responsibility during view periods was to maintain both uplink and downlink communications to ensure the astronauts' safety and the mission's success. Shortly before the Apollo launch, on 16 July 1969, NASA's worldwide tracking network received a briefing from the Goddard Space Flight Centre in Greenbelt, Maryland. Perhaps with the example of the mop handle in mind, the overall network controller launched forth:

> The chances of hardware problems in the spacecraft and on the ground which could seriously jeopardise the mission's success were much less than the chances of a person pushing the wrong button at the wrong time. For example, unless the antenna is pointed at the right place at the right time, the station might as well not be there. Also, when the antenna technician has done his part, the transmitter and receiver technician must push the right buttons at the right times if any data or voice up or down to the spacecraft is to be received. This operational performance follows down the line, i.e. the chain is as strong as its weakest link.[14]

The Upside-down Camera

Unlike most of his predecessors, Apollo 10's commander, Tom Stafford, had been a TV enthusiast. It was largely thanks to him that the first live colour television broadcast of an Apollo mission had occurred. During Apollo 10's journey to the Moon, audiences the world over had watched in awe as the Earth and by extension the whole of human history, appeared to shrink on their TV screens. Of this increasingly distant, brilliantly coloured orb, cosmologist Carl Sagan later said:

> On it … every human being who ever lived, lived out their lives … every hunter and forager, every hero and coward, every creator and destroyer of civilizations, every king and peasant, every young couple in love, every hopeful child, every mother and father, every inventor and explorer, every teacher of morals, every corrupt politician, every superstar, every supreme leader, every saint and sinner.[1]

By simply holding up his thumb, an astronaut could completely hide the Earth from sight. There was also live colour footage of the command and lunar modules manoeuvring near each other, together with some panning shots of the Moon's surface, including close-ups of the next mission's proposed landing site on the Sea of Tranquility. All this filming was done from Apollo 10's command module *Charlie Brown*. There was no television camera aboard its lunar module *Snoopy*. And it would be some weeks before a final decision would be made whether or not to attempt a live television broadcast from Apollo 11's lunar module *Eagle*. For despite a report in the *Canberra Times* of 15 May 1969 that 'under present plans ... a TV camera will be erected on the Moon's surface by the astronauts to record their activities,' NASA was still undecided about whether this was either necessary or desirable.[2]

As early as April 1961, the grandly named Apollo Technical Liaison Group for Instrumentation had almost killed off the idea of taking TV cameras to the Moon, by pointing out that the space needed on a downlink to carry high-quality television signals back to Earth would swamp the space available for crucial voice and telemetry signals. But NASA's public relations department had fought back, arguing that as the stupendous cost of the Apollo program was being underwritten by ordinary American taxpayers, the public should be able to see the first human being step onto the Moon – on live TV from the comfort of their living rooms. From mid-1961 until just weeks before the Apollo 11 launch in July 1969, debate raged between NASA's pro-TV and anti-TV factions.[3]

The NASA faction which supported live lunar TV broadcasts had soon realised that any attempt to do so with conventional television equipment would be impossible. Given the likely weight, space and bandwidth constraints on the spacecraft that would attempt a lunar landing, revolutionary new TV technology would be needed. Without committing itself to live TV from the Moon, NASA resolved to fund the technological innovations necessary to give it the option to do so.[4]

NASA also increased its options for such broadcasts by upgrading its Australian tracking network. When added to the existing capabilities of Goldstone and the 85-foot dish at Madrid, this would for the first time allow live lunar TV coverage around the clock. Because Goldstone's 210-foot dish had delivered the best television pictures from Apollo 8 and 10, NASA designated

Parkes's similar sized dish as the prime Australian receiver of TV signals from the Moon. To equip Parkes and bring it up to speed at short notice, a NASA team from the States headed by Goddard's Robert Taylor was embedded there. While Parkes's other downlink data would be sent to Honeysuckle as previously planned, its TV signals would be forwarded to Sydney, where after being processed, they would be transmitted to Houston.[5]

If NASA was going to publicly commit to live TV of Apollo 11's astronauts walking on the Moon, it needed to have TV redundancy at each of its key tracking stations. In Australia's case, if Parkes's big dish was unable to receive the lunar module's downlinked TV signals for any reason or if its picture was faulty, Tidbinbilla rather than Honeysuckle was to be its primary back-up. This was because Tidbinbilla's more sensitive receiver was allocated to the lunar module's downlink, while Honeysuckle's was locked onto the command module. Even so Tidbinbilla's TV signals were, like all the rest of its downlink data, to be sent on to Honeysuckle for processing and then for transmission to Houston. In practice Houston was to receive TV pictures from the various dishes that had the lunar module in view at any particular time and then select the best live feed.[6]

To deal with downlinked TV signals, Honeysuckle received an urgent upgrade. At that time the standard American TV network format involved a scan of 525 lines and thirty frames per second. But the lunar module's downlink did not have sufficient bandwidth to carry this format. The most it could handle was a scan of 320 lines and ten frames per second. So rather than presenting the public with a jerky and unintelligible TV show, NASA had commissioned the RCA Corporation to develop a scan converter which crudely described, involved putting the slow scan picture on a cathode ray tube and then pointing an ordinary TV camera at it. Goldstone and Madrid already had such converters for their dishes while Parkes's were to be processed through a NASA converter in Sydney. A converter identical to these was now added to Honeysuckle's telemetry operations area, primarily to process

TV signals coming in from Tidbinbilla. But if for whatever reason the tracking roles of Tidbinbilla and Honeysuckle were switched, this converter could handle any TV signals coming in via Honeysuckle's own dish too.[7]

A transmission link from Honeysuckle to Canberra was built to get the converted commercial grade TV signal through to Sydney and then on to Houston. A back-up link was also needed and a special team hastily cobbled together various surplus small dish antennas using a lathe out the back of Honeysuckle's operations building. These were placed at the top of a temporary tower erected beside the road into Honeysuckle, together with linking equipment borrowed from the ABC. From this tower, the signal would be sent to the small village of Williamsdale on the Monaro Highway and then to a microwave relay station on top of Canberra's Red Hill. One of the installation team recalled:

> These links were difficult to maintain – it was wintertime and very cold. There was a high voltage in these joints and moisture got in them and a few of them blew up...We ended up sealing them up with epoxy resin. Normally these links were designed to be only up for an hour or so, say during a football match; they weren't meant to stay there that long.[8]

Despite all this extra equipment in place, the debate between NASA's pro-TV and anti-TV factions raged on. In June 1969 it all came to a head when the Apollo communications' director, Ed Fendell, flatly recommended against taking TV to the Moon. NASA's director of flight operations, the magnificently named Christopher Columbus Kraft, then weighed in. And he won the day by forcefully putting the time-worn argument that American taxpayers had the right to observe the mission their taxes were financing. But with just a couple of weeks to go until the commencement of the Apollo 11 countdown, there was very little time left to install the light-weight black-and-white TV camera which had been specially designed for NASA, just in case lunar

TV was given the go ahead. This camera incorporated a low-light television imaging tube developed by the United States Defence Department to locate its downed pilots at night in North Vietnam. The tube was classified top secret.[9]

Implicit in the camera's design was the proposition that nothing less than an outside television broadcast of the first human being actually setting foot on the Moon's surface would do. The identity of the person most likely, had been settled on 6 January 1969 when NASA's senior astronaut, Deke Slayton, had called Neil Armstrong, Edwin 'Buzz' Aldrin and Michael Collins into his office. 'You're it guys,' Slayton said. 'You've got the Apollo 11 flight, and that means you get first crack at landing on the Moon. That is, of course if we pull off successful missions with 9 and 10.' While Collins would remain in charge of the command module, *Columbia*, during the mission's critical lunar phase, Armstrong and Aldrin would attempt a Moon landing in the lunar module *Eagle*, to be followed by a Moon walk. 'Just on a pure protocol basis,' Slayton said, 'I figured the commander ought to be the first guy out.' While Slayton's comments appeared offhand, he knew that if Apollo 11 was a success, Neil Armstrong's name would be remembered down the ages as the first human to set foot on a celestial body. And the question was how to capture his first step for live TV.[10]

With no atmosphere on the Moon's surface to allow the *Eagle* to fly in the traditional sense, it was the world's first true spacecraft, relying on the firing of its descent stage rockets for a survivable landing on four spindly, strutted, telescopic legs. Later, following the astronauts' Moon walk, this descent stage would form the launch platform for the top half of the *Eagle* to blast off back to the *Columbia*. For both the descent and ascent stages, keeping weight to a minimum was of crucial importance. The lunar module's external aluminium and mylar skin was therefore paper thin. According to Gene Kranz, 'You could easily poke a pencil through *Eagle's* side.' To further reduce weight, the astronauts' seats were dispensed with; tethered in place with straps, Armstrong and

Aldrin would fly standing up. With these sorts of weight limitations in mind, it was a Westinghouse engineer, Stan Lebar, and a team of seventy-five, who developed a special TV camera weighing just six pounds at a time when a conventional camera weighed 400 pounds. Shaped like a small, slope-sided baking dish with its lens protruding from one end, Lebar's camera could be set up by the astronauts on the lunar surface a short distance away from the *Eagle*. A marvel of electrical engineering, this little camera could operate on just six watts of power. Yet it had a low light capability, could withstand anywhere between −157°C and +121°C (−250 and +250° F) and could be automatically operated. Connected to the lunar module by a TV cable, the camera would be able to remotely film much of the astronauts' Moon walk. Its images would be transmitted live through the cable to a small S-band antenna on top of the *Eagle* and then on its downlink signal to whichever tracking stations happened to be in view.[11]

Televising the first step presented special problems in as much as NASA, having now committed to it, literally wanted live footage of Armstrong stepping off the *Eagle's* external ladder and onto the Moon's surface. There was a lot riding on the technology involved. While a successfully televised first step would be the ultimate public relations coup for the United States, a botched attempt would be far worse than no attempt at all. And the pressure on Stan Lebar to get it right was intense. Attached to the top of one of the landing legs was a platform which the astronauts called a 'porch'. NASA's original plan was for Armstrong to crawl through the *Eagle's* external hatch and stand up on the porch. Aldrin would then pass him a TV camera to take a panning shot of the lunar surface. That done, Armstrong would hand the camera back. And Aldrin would film him climbing down a ladder that was attached to a landing strut and onto the Moon. But NASA's simulations showed that there was a high risk of Armstrong overbalancing, falling off the porch and likely being killed. So that plan was abandoned in favour of a remotely triggered and operated camera which would need to capture

Armstrong climbing down, negotiating a gap between the bottom of the ladder and a dish-shaped landing pad and then stepping onto the lunar surface.[12]

After much discussion and experimentation, Stan Lebar and the *Eagle's* designers settled upon storing the TV camera in the descent stage's modular equipment storage assembly or MESA. In plain English, the MESA was a tool locker which would automatically open outwards after Armstrong had emerged onto the *Eagle's* porch and pulled on a special lanyard. Lebar's camera was mounted in such a way that when the MESA was opened by Armstrong, the lens would be pointing straight at the *Eagle's* ladder, ready to film his first step. There was a complication. Because of the tight space inside the MESA, there was only one way to mount the camera. And this particular way meant that the camera would be upside down when it filmed the ladder. On all the tracking stations' scan converters, a special toggle switch was incorporated to flip the upside down TV images being received from the *Eagle*, right side up. With these systems in place, Lebar's little camera was subjected to a series of brutal water, vibration, dust and sealant tests. Unlike most of Apollo 11's other systems the camera had no back-up. If it didn't work there would be no second chance. As an article from Westinghouse's in-house magazine put it: 'The final test of the lunar TV camera will be viewed by millions as they watch on their home television sets.'[13]

According to the Apollo 11 flight plan, which had been signed off on 1 July 1969 by NASA's big wigs, the Australian tracking stations were slated to play a key role. This was summarised in a press release issued by Honeysuckle's public relations manager, Bernie Scrivener. Among other things Scrivener described what was scheduled for 4.12 pm on 21 July, Australian eastern standard time:

> Man, in the person of astronaut Armstrong, emerges to set foot on the lunar surface. During this period of some two (2) hours forty (40) minutes of lunar surface activity the Canberra

> complex, joined by their colleagues at the CSIRO 210 ft antenna, Parkes, will be the prime and only links between Mission Control at Houston and the astronauts on and around the Moon.[14]

Coordinated by Tom Reid, Honeysuckle, Tidbinbilla and Parkes and their combined staff of almost 200 would be responsible for maintaining all communications with Apollo 11 during the crucial Moon walk phase. These communications included the televising of Armstrong's first step and the monitoring of his heart beat and respiration rate as he did so, while at the same time enabling him to give Mission Control a second-by-second description of his progress. At Honeysuckle the lead up to launch day passed in a blur of twelve-hour shifts over sixteen straight days. The station's teletype machines went into overdrive, clattering away at all hours of the day and night as they churned out messages, reports and instructions relating to administration, station status, operational updates, engineering changes, mission countdown status and press briefings. As Bryan Sullivan recalled:

> The teletype operators delivered endless wads of paper to every section of the station. Documentation clerks cut and pasted the latest updates into the many copies of an extraordinary number of operational documents, from the mission flight plan down to the local equipment operating procedures.[15]

The intensity of all types of simulations were also stepped up: simulations involving Mission Control, Goddard and the worldwide tracking network; simulations limited to the Australian tracking complex which now included Sydney Video where a senior NASA staffer, Charlie Goodman, would select the best of the lunar TV feeds coming in from Honeysuckle, Tidbinbilla and Parkes for on-forwarding to Houston; and simulations focused on Honeysuckle itself. Sometimes there were unintended consequences. 'We actually managed to tie up most of the communications around the East coast of Australia when we had an internal simulation,' John Saxon said. 'Channel 7 didn't always get their news at the right time because we had all the television feeds tied up and they had to delay the evening news items.'[16]

Following a thorough check of all systems, including a green light roll call of Goldstone, Madrid and Honeysuckle, the final countdown at Cape Kennedy began. And on the morning of 16 July 1969, in front of a crowd of over one million people who crammed onto the local beaches, camping areas and parking lots, Apollo 11 blasted off. With a maximum thrust equal to 180 million horsepower, Wernher von Braun's massive Saturn V rocket carried the Moon-bound spacecraft into Earth orbit, after which the astronauts onboard *Columbia* and with the *Eagle* attached, altered course for the Moon. When the spacecraft had travelled about 10000 miles, Goldstone, Madrid and Honeysuckle, one after the other, began their collective 24-hour track.[17]

All went well for two days. Then disaster struck at Tidbinbilla. As that station's transmitter field engineer, Alan Blake, later recalled:

> I had just come down from dinner in the canteen when there was a call on the intercom to say there was a fire in the transmitter power supply. I ran out there and found dense, thick black smoke was everywhere, and knocked off the circuit breaker and waited for the smoke to clear, then went into the cabinet. There was a horrible mess in there. The temperature had been so hot that the top of the cabinet had buckled. It took quite some time for it to cool down enough to touch it.

Faced with what would normally have taken a week to fix, Alan Blake and his team laboured like men possessed to work up a replacement system. After chopping out the old chunks of cable loom, new bunches of wires were made up and fitted. At the same time, all the removeable units were pulled out and a steady stream of technicians ran down to the station's store with them and any technical drawings they could find, to get replacement parts. 'We used anything and everything that could fit in,' Blake said. Although NASA was grateful for these 'heroic efforts' which had Tidbinbilla back on line within twelve hours, its engineers were concerned that the repairers had been forced to improvise by scrounging whatever spare parts were available. Such repairs were known in the trade as a 'kluge', the problem being that no one knew how reliable they might be. This was of great concern because Tidbinbilla was scheduled to handle the communication links not just with the *Eagle* but also with Armstrong and Aldrin via their backpacks while they were walking around outside. As one of Houston's network controllers Richard Starchurski noted, the risk was just too great 'to trust the crew members to the jury-rigged equipment'. After urgent top-level discussions between Houston, Goddard, Tom Reid and Don Gray, it was decided that Honeysuckle's slightly weaker receiver would handle the uplink and downlink with the lunar module, and by extension the astronauts during their Moon walk, while Tidbinbilla's jury-rigged dish would look after the command module. In relation to live TV broadcasts from the Moon, this also meant that Honeysuckle's 85-foot dish would now be the prime back-up for the much bigger dish at Parkes. If for whatever reason Parkes was unable to receive the *Eagle's* live TV signal, it would fall to the little dish at Honeysuckle to bring Armstrong's first step to the world. While all was frenetic activity at Honeysuckle and Tidbinbilla as the stations' staff raced to get across the switch in their respective responsibilities, Apollo 11 voyaged on serenely towards the Moon, with Armstrong and Aldrin scheduled to make a lunar landing attempt early on the morning of 21 July, Eastern Australian time.[18]

Like Flies to a Picnic Lunch

As a teenager in the Royal Navy during the mid-1940s, Tom Reid had learned how to snatch sleep in awkward places. Curled up in a hammock swinging to the rise and fall of his warship at sea, he'd had to contend with the sounds of machinery and of men, many of them snoring, while others scrambled about on watch. At times HMS *Victorious* had seemed like a giant living thing that never rested.

Now forty-two, Tom was again trying to sleep, this time in the basement of the Honeysuckle Creek tracking station which during manned missions to the Moon never rested either. Against the background hum of an air-conditioning plant powerful enough to prevent overheating in a building stuffed full of the latest computers, teletype machines, switching gear, consoles, communication loops and bank after bank of monitors, and across the way from a full-service canteen which operated twenty-four hours a day during Apollo missions, he did eventually manage to catnap.

Awake again by 4 am, Tom lay in the bunk of his windowless bedroom, drawing deeply on a cigarette as he contemplated the day ahead: Monday 21 July 1969. Although by nature and training a hard-driving technocrat who suppressed any sentimentality, his first thought this day was of how as a wee lad he had told his mother that during his lifetime a man would walk on the Moon. Recalling her incredulous response, he chuckled to himself; if all went according to plan today, his mother would be watching from the comfort of her Launceston living room as Neil Armstrong stepped onto the Moon at around 4.30 pm.[1]

Having been involved in space tracking for longer than most people, Tom felt quietly confident that today's Moon landing would be a success. Apollo 11 was the culmination of a stupendous effort by a worldwide team of 400 000 men and women who over the previous ten years had designed and built a series of rockets and spacecraft, each one more complex than the one which had preceded it, firstly to put a man into space; then to put a man into orbit; then to conduct spacewalks; then to dock two separate spacecraft; then to fly out to the Moon, into Lunar orbit, and return to Earth; and then, most recently, to fly a lunar module to within 50 000 feet of the Moon's surface. The rockets and spacecraft, moreover, had been just the tip of a giant technological iceberg which among many other things had included the development of rocket fuel, space suits, launch pads and computers. There had also been the myriad calculations necessary to plot the course of spacecraft which at various times would have to alternately defy and take advantage of the gravity of the Moon and of the Earth, especially the extreme heat of re-entry, all the while allowing for the fact that the Moon circled the Earth which in turn revolved on its own axis once every twenty-four hours. Reflecting on all this, Tom was momentarily awed by the thought that the vital voice, command and telemetry links to Armstrong and Aldrin as they walked upon the Moon, not to mention the live TV feed, would be his responsibility as the overall coordinator of Honeysuckle, Tidbinbilla and Parkes.[2]

While finishing his cigarette, Tom Reid had one final thought before swinging his legs onto the bedroom floor and heading off down the corridor for a shower: that in the last six months, as the Americans had upped the tempo of missions with Apollo 8, 9 and 10 seemingly happening within just weeks of each other,

Honeysuckle had performed magnificently. While the Apollo 10 lunar module had been descending to within 50000 feet of the Moon's surface less than eight weeks ago, massive volumes of instructions for Apollo 11, specially flown out from Houston, had thumped down on every desk and console in Honeysuckle's operations area. And in the short time since then, every engineer and technician in Tom's team had become thoroughly familiar with their new briefs. In rebuilding and readjusting after the recent fire at Tidbinbilla, those he coordinated had also demonstrated just how calm and innovative they could be in a crisis. While he was damned proud of them all, he was careful not to let it show.[3]

Late the previous day he had received a call from his boss, Lloyd Bott, the deputy-secretary of the Department of Supply and a great supporter of Honeysuckle, to warn him that a VIP was intent on visiting the station around 8.45 am – the only VIP in the country who could not be put off. Keen to listen uninterrupted in his office to the various communication nets which would allow him to monitor the lunar landing scheduled to commence shortly after 5 am, and keen to be out and about on the main operations floor to observe how the station was being made ready for its view period scheduled to commence around 11.15 am, and it now being 4.30 am, Tom decided to get cleaned up straight away. On most working days he would put on a white shirt and tie like the other station staff. But because he was staying over, he had brought more casual clothes and did not have time to get home to Hughes for the suit he usually wore during VIP visits. After shaving and showering, Tom put on what he'd brought with him: slacks, an open-necked shirt, a cardigan and a polka-dot cravat which would just have to do for this particular VIP. Although by 1969 wearing a cravat was unusual in Australia, it was a personal style derived from the scarves Tom and his father had worn on winter Sunday outings in Glasgow during the Great Depression.[4]

Having made himself presentable, Tom Reid ducked across the corridor to the station's canteen. There Betty and Horrie Clissold were serving up hearty breakfasts of bacon and eggs to a hungry night shift. The engineers and technicians waiting in line knew that this was their last chance to grab a bite before Honeysuckle's demanding station readiness testing began at 6 am. Although the station's catering was legendary and attracted staff from Tidbinbilla 25 miles away, Tom wasn't tempted, not just now anyway. Instead, he settled for a large mug of strong black coffee. Occasionally when Tom and his operations manager, John Saxon, had worked overnight at the station in the past, they had added a dash of Bourbon as a pick-me-up first thing in the morning. While alcohol was prohibited at the American-run tracking stations, which like the United States Navy were dry, the navies Tom had served in followed a different tradition. However, today wasn't the day for turning a blind eye to even the tiniest nip. And he settled for another cigarette instead.[5]

After moving upstairs to the station's main operations area, Tom walked over to the only window on this floor which was located immediately above a console manned by the technician who 'drove the dish'. From this position the station's main, 85-foot, concave antenna could be moved vertically or horizontally, or in combination, at a maximum rate of 5 degrees per second – fast enough to knock a person jogging around its base off their feet. Although the window was there so the operator could see what he was doing when moving the dish at speed, the reality was that its computer-generated movements when tracking a spacecraft were barely perceptible to the human eye.[6]

While it was still pitch black in the surrounding mountains with the sun not expected to rise until just after 7 am, the dish itself was brilliantly illuminated. Whitish in colour, it cast its reflected artificial light over a wide area. This allowed Tom to see the kangaroos which grazed on the campus-like grass immediately beyond the dish's concrete base. This morning he was particularly interested in the weather. Although there was little wind, it seemed bitterly cold with frost making the kangaroos' foraging difficult. There also appeared to be a relatively low cloud cover. With snow expected above 4000 feet, Tom wondered what the chances were of snow falling into the dish, the apex of which in a vertical position, topped 3800 feet above sea level. In these mountains you just never knew. Indeed a few weeks earlier while tracking Apollo 10, Honeysuckle's dish had had to be tipped right over to empty it of snow. And if there was a heavy fall today, the station would again have to 'break track' to get rid of the stuff. Outside, anyone not wearing gloves would leave skin on the handrails. Still, Tom was comforted by the fact that the station's maintenance team led by a fellow Scottish engineer, Jim Kirkpatrick, was more than up to the job of de-icing the dish and otherwise keeping its mechanical, hydraulic and electrical systems from seizing up in the cold.[7]

Despite the snowfall during Apollo 10, it was still difficult to get NASA to understand the tracking challenges presented by a Canberra winter. In the northern hemisphere where the other two tracking stations were located, it was high summer. At Goldstone in California's Mojave Desert, the temperature sometimes topped 100 degrees while in Madrid it could get even hotter. Although a number of NASA officials had visited Honeysuckle, they tended to do so during Australia's warmer months. Besides most Americans simply could not envisage snow in Australia. To them it was a wide brown land of endless surf and sunshine. And in winter, Houston's climate was more like Sydney's than Canberra's. At Honeysuckle, this was known as the 'Southern Hemisphere problem'; NASA tended to think only in terms of whatever was happening north of the Equator. A wry smile played on Tom's lips as he contemplated how he would inform Mission Control in sultry Houston if Honeysuckle had to again break track to dump a dish full of snow. Perhaps for once his Glaswegian accent might help.[8]

Having surveyed the weather as best he could through the window, Tom moved quickly to the operations console at the other end of the building. There one of his two deputy directors, Ian 'Iggy' Grant, another Scot, was seated in the Ops 1 chair with Ken Lee beside him. Between them they seemed to have the upcoming station readiness tests well in hand. Satisfied, Tom strode down the hall to his office, the door to which had painted onto it in large capital letters 'Station Director Private'. Below this in a smaller font were the words 'Enquiries with Secretary'. Furnished with the sort of non-descript desk, chairs and bookcases typical of the Commonwealth public service, the only concession to Tom's seniority was that they were all made out of well-polished wood. Apart from some bookended briefing folders in various blue/grey colours, the only other adornments were a couple of anaemic pot-plants, two modestly framed colour photos of previous missions and a model rocket. As always, Tom's desk was sparsely covered: by a blotter, a ruler, some writing implements and the obligatory ash tray. Apart from the sign on the door, the only things to suggest that an important person occupied this room were a large floor safe; the shiny combination padlocks and metal bars that provided extra security for a series of already locked filing cabinet drawers; a phone which unusually for that era could make direct calls to the United States and to Spain among other places; and a special apparatus that allowed Tom to listen in on NASA's various communication links. These included Net 1, which carried conversations between Mission Control and the astronauts; Net 2, which allowed Mission Control's network controller to coordinate Goldstone, Honeysuckle and Madrid; and Net 3, which enabled NASA to double-check an individual tracking station's readiness as its view period approached. Tom had stayed over at the station this past night, in part so he could hear Mission Control over these nets during the *Eagle's* landing phase. Although much of what would be said would be broadcast by Canberra's radio stations, the listening public would only hear what NASA wanted them to hear. And it was the bits that NASA was likely to hold back that Tom was most intent on listening to.[9]

From his previous visits to Houston, Tom could picture the control room on the third floor of what was known as Mission Control Centre in Houston. Here in a tiered theatre-style arrangement, four rows of controllers faced a giant bank of screens. Behind these rows in a sealed off observation area were NASA's luminaries, among them its chief administrator Dr Thomas Paine, its chief rocketeer Dr Wernher von Braun and its most famous astronaut John Glenn. Seated in the middle of the back or top row, closest to these sealed off luminaries, was the director of flight operations, Chris Kraft, known as DFO. To his immediate left was NASA's public affairs officer; between them they would determine just how much of today's proceedings would be broadcast to the public. Immediately below, in the middle of the next row down, was the flight director, Gene Kranz, known as FLIGHT, with the tracking stations' overall controller, George Ojalehto, known as NETWORK, sitting two seats to his right. Among those in the row below them was a surgeon, next to whom sat CAPCOM (the astronaut Charlie Duke) who except in an extreme emergency, was the only person in the room or anywhere else in NASA's worldwide network authorised to talk to the Apollo 11 astronauts. And in the front row, affectionately known as 'The Trench', were the engineers in charge of boosters, retro-firing, flight dynamics and guidance. Each of these controllers was supported by specialists sitting at their consoles in rooms adjacent to the main control room.[10]

For Mission Control purposes, Apollo 11 was divided up into phases, each one of which was presided over by a flight director and his team who filled the front three rows of the control room during their particular shift. The green team whose flight director was Clifford Charlesworth, had handled the launch a few days earlier and would also be handling the Moon walk later today; Gene Kranz and his white team were about to handle the lunar landing; and tomorrow, Glynn Lunney and his black team would cover the lunar module's take-off from the Moon. These teams were responsible for monitoring the astronauts' life support systems and vital signs, together with all systems aboard the command and lunar modules; sending these spacecraft remote commands; talking as necessary with the astronauts to provide advice and help solve problems; and most importantly, determining whether or not to 'Go' with the next stage of the mission, such as the lunar landing or the Moon walk. To carry out their vital work, they relied completely on the command, voice and data signals, which were being sent backwards and forwards between their consoles and the Apollo 11 astronauts, via Honeysuckle Creek, Goldstone or Madrid, depending which of these three tracking stations at any particular time had the Moon in view.[11]

Almost thirty-six, Gene Kranz was a former fighter pilot who, after serving in Korea, had been with NASA for almost nine years. With a buzz cut which made the top of his head as level as an aircraft carrier's flight deck and an embroidered white vest crafted specially by his wife to symbolise his leadership of the white team, the hawkish-looking Kranz cut a flamboyant and dominating figure. In some ways he resembled the conductor of an orchestra, even though he sat behind and above those he conducted. Looking intently at their consoles which were arranged in tiered rows below, Kranz's team communicated with him through voice loop headsets. As Tom Reid listened over NASA's communication net to the goings on at Mission Control sometime shortly before 6 am, he heard Kranz's call: 'Ground Control, lock the control room doors.' No controller could enter or leave the room. Then as Tom's pulse quickened, he heard Kranz bark: 'Take Mission Control to battle-short.' A phrase familiar to Tom since his wartime naval service, it allowed equipment to bypass circuit breakers which normally protected it, so it could be kept continuously operating in combat. By now, the command and lunar modules had separated and each was independently orbiting the Moon. At this time, the tracking stations at Goldstone and Madrid both had the Moon in view, in so called 'mutual view'. As such they were each able to communicate with the astronauts when they were orbiting around the near side of the Moon – the side which faced the Earth. While Goldstone now tracked Armstrong and Aldrin in the lunar module, Madrid tracked their colleague, Michael Collins, in the command module.[12]

After achieving independent flight, Armstrong and Aldrin communicated with Mission Control via Goldstone, using a 26-inch diameter parabolic dish which was mounted on top of their lunar module, near its access hatch. This tiny dish was designed to be operated manually until such time as it could lock onto Goldstone's uplink signal, after which it was supposed to track automatically. As such, its performance was of vital interest to Tom Reid who knew that later that same day his Honeysuckle dish would have to lock onto it for the all-important Moon walk. What Tom was especially waiting to hear was for the acquisition of signal, 'AOS', between Goldstone and the lunar module, and for any loss of signal, 'LOS', between them. While Mission Control was invariably referred to over the voice nets as 'Houston' and the command module was called *Columbia*, the lunar module was sometimes called 'LEM' and at other times *Eagle*.[13]

With a fresh cigarette and what was left of his coffee, Tom listened intently for the no-nonsense exchanges which would soon flow backwards and forwards over the communication nets between Houston, *Columbia* and *Eagle*. For the last little while, the command and lunar modules had been on the far side of the Moon and therefore beyond contact with Houston. During that time, Armstrong and Aldrin had executed a descent orbit injection burn, an important precursor to their landing which slowed the lunar module's orbital velocity, allowing the Moon's gravity to pull the *Eagle* towards its surface. Like everyone at Houston, Tom waited with bated breath to hear from Michael Collins in the command module, which would be the first of the spacecraft to re-establish contact with Houston. What Houston wanted to know from Collins was whether or not this crucial burn had been a success. Before long, Madrid picked up the command module's signal. But there was no word on how the burn had gone. Unable to stand the tension any longer, CapCom Charlie Duke, pushed for a report:

'*COLUMBIA*, HOUSTON. We're standing by. Over.'

'*COLUMBIA*, HOUSTON. Over.'

'HOUSTON, *COLUMBIA*. Reading you loud and clear.'

'Roger. Five-by (OK), Mike. How did it go? Over.'

'Listen, babe. Everything's going just swimmingly. Beautiful'...

'GOLDSTONE has LEM AOS'...

'HOUSTON, *EAGLE*. How do you read?'

'Five-by, *EAGLE*. We're standing by for your burn report, over.'

'Roger. The burn was on time'...

'GOLDSTONE has LOS LEM.'

'*COLUMBIA*, HOUSTON. We've lost all data with *EAGLE*.'

For what seemed like an eternity, Tom listened intently to this most critical phase of the Apollo 11 mission as Houston reacquired, lost and reacquired voice contact with Armstrong and Aldrin on at least three separate occasions. 'We did not have to look for problems because they came right at us, like flies drawn to a picnic lunch,' Gene Kranz later recalled. 'Voice communications were broken, and LM telemetry was unable to lock up. The noise on the air-to-ground communications loop was deafening.' Along with everyone else, Tom tried to work out what was going wrong. Like the tracking stations, Houston and the astronauts had been subject to simulation tests while preparing for Apollo 11, including total loss of communication. But no one had predicted or planned for the intermittent communications being experienced. 'At times the intercom crackled,' Kranz said, 'The voices diced and chopped, staccato.' It was a most disturbing development and Tom wondered how he would handle this problem should it occur on his watch.[14]

The *Eagle* Has Landed

As Gene Kranz faced the mission's most crucial question, whether it would it be 'Go' or 'No Go' for a lunar landing, Houston's communications with the *Eagle* kept breaking up. He and his white team did not have the constant stream of vital telemetry needed to advise them. Kranz's flight dynamics and retrofire officers, FIDO and RETRO, struggled to get accurate information on trajectory; his guidance officer, GUIDO, was having trouble monitoring the navigational software; his control officer, CONTROL, could not follow the *Eagle's* propulsion, guidance and navigation systems; his telemetry officer, TELCOM, was unable to accurately measure its electrical and environmental systems; and the surgeon struggled to follow the astronauts' respiration and heart rates. With just minutes to go, Kranz said a prayer: 'Please God, give us comm.' Right on cue, there was a brief burst of telemetry from the *Eagle*, sufficient for Kranz to poll his controllers:

'RETRO? GO.'

'FIDO? GO.'

'GUIDO? GO.'

'CONTROL? GO.'

'TELCOM? GO.'

'SURGEON? GO.'[1]

Satisfied, Kranz called Charlie Duke: 'CAPCOM, FLIGHT. We're GO for powered descent.' Then NETWORK, George Ojalehto, came on line: 'LOS for the LEM downlink, FLIGHT.' Houston's connection to the *Eagle* had dropped out. Even so Charlie Duke made an attempt to get through: '*EAGLE*, HOUSTON. If you read, you're GO for powered descent. Over.' But there was no response. Michael Collins, meanwhile, had been monitoring what Duke was saying and so he radioed Armstrong and Aldrin: '*EAGLE*, this is *COLUMBIA*. They just gave you a GO for powered descent.' Although it was not immediately clear that the *Eagle* had received this message, its trajectory began to change as if it had. And as it commenced to straighten up relative to the Moon's surface, intermittent communications were re-established. As Gene Kranz later recalled: 'Radar continued to be ratty and was frequently lost.' With the next 'Go or No Go' looming, Kranz came back on line. 'OK, all flight controllers, I'm going to go around the horn. Make your GO/NO-GOs based on the data you had prior to LOS.' When they had all responded with a 'GO', Charlie Duke made the most of a sudden reconnection: '*EAGLE*, you are GO – you are GO to continue powered descent. You are GO to continue powered descent.'[2]

No sooner had Armstrong acknowledged this message than a loud alarm went off in the lunar module. Neither he nor Aldrin had any idea what it meant. And there was no time for them to rifle through their onboard flight manual to find out. Armstrong pressed Houston for advice: 'Give us a reading on the 1202 PROGRAM ALARM.' It was in fact a computer signalling an overload. If it was a genuine alarm, the Moon landing would have to be aborted. Houston's guidance officer, Steve Bales, was in the hot seat with Gene Kranz demanding to know whether or not it was a false alarm. Unsure of the answer, Bales called his support team in the adjacent room for urgent advice. Luckily a twenty-four-year-old spacecraft computer specialist, Jack Garman, could recall a simulation where as long as this alarm was intermittent, the descent could continue. And so, it did: '*EAGLE*, CAPCOM. We're GO on that alarm.' After ignoring further 1202 alarms which were so close together that for a moment they sounded like one dreaded continuous alarm, the *Eagle* was soon within 5000 feet of the Moon's surface. The quality of the voice and telemetry data was now 'close to perfect'. Kranz again polled his team: 'OK, all flight controllers, GO/NO-GO for landing?'

'RETRO? GO.'

'FIDO? GO.'

'GUIDO? GO.'

'CONTROL? GO.'

'TELCOM? GO.'

'SURGEON? GO.'[3]

Armstrong, meanwhile, had realised that due to a slight navigational error, he had overshot his intended landing place by 4 miles and couldn't rely on the Eagle's pre-programmed computer to guide the touch down. Making matters worse, he was looking at a landing field littered with boulders that had not shown up on the photos taken by Apollo 10 a few weeks earlier. What the worldwide radio audience of hundreds of millions now heard, was Armstrong manually piloting the *Eagle* down towards and across the lunar surface looking for another spot, as Aldrin called out their rate of descent. The astronauts sounded so cool under pressure that Tom Reid was one of the few who understood how tight their margin for error was. From the incoming reports, Tom knew that the Eagle's fuel levels had fallen below the point where they could be accurately measured, as Armstrong dodged craters and boulders searching for a level landing site. It was as if Armstrong's fuel gauge was on empty at a time when the lunar module had entered the 'dead man's curve', flying too low to abort if the engine conked out. After what seemed like an eternity, Tom heard Armstrong say: 'HOUSTON, TRANQUILITY BASE here. The EAGLE has landed.' As a hard-bitten Glaswegian, Tom did not believe in heroes but what he had just been monitoring was more than coolness under pressure; it was a display of great courage.[4]

Taking a few moments to reflect on how what he had just heard might impact on Honeysuckle's planned tracking later in the day, Tom Reid was struck by the way the balance of power between Gene Kranz and Neil Armstrong had altered during the final moments of the lunar module's descent. Knowing that Houston had been unable to do any more to assist Armstrong with his landing, Charlie Duke had told Kranz over their internal voice loop: 'I think we'd better be quiet.' As Kranz himself later acknowledged, at that point 'the centre of gravity of the decision-making process' had moved from being at some imaginary point in space, halfway between Houston and the *Eagle*, to being solely with Neil Armstrong. While this power shift to Armstrong would inevitably rebalance back towards Houston as the mission progressed, Tom knew that Armstrong would have a much greater say than originally planned in what happened during the next part of the mission – the Moon walk.[5]

Also playing on Tom's mind were the intermittent communication failures between Houston and the lunar module which had plagued the most critical moments of its descent to the Moon. From listening in on Nets 1 and 2, Tom was aware of the theories: that somehow Goldstone had been unable to lock onto the *Eagle*'s main downlink beam, instead connecting up with one of its low-powered side beams that did not have enough power to carry the necessary telemetry; that as the *Eagle* had altered its flight path to close in on the Moon's surface, its steerable antenna had been pointing too close to part of the spacecraft's structure, thereby causing the antenna's signal to distort; or that the *Eagle* had been flying at an angle that made it difficult for its antenna to remain pointed at Goldstone. Whatever the case, Tom took comfort from the fact that as the lunar module had straightened up for its final approach, its signal had markedly improved. As Mission Control and Armstrong now worked their way through the post-landing decision of whether to stay or go, with the *Eagle* configured so that if necessary, its ascent stage could take off immediately, the 284 000-mile-long electromagnetic ribbon that connected them via Goldstone operated faultlessly. Within minutes the verdict was in: '*EAGLE*, CAPCOM. You have a GO for extended surface operations'; confirmation to Tom Reid that by late afternoon, Honeysuckle would be the prime communication link between Houston and the first person to set foot on the Moon.[6]

Having drifted off course just prior to landing, Armstrong and Aldrin could not pinpoint their location. As Armstrong put it, 'The guys who said we wouldn't know where we were are the winners today.' Unable to recognise any nearby landmarks, Armstrong left the job of locating their precise whereabouts to Houston while he and Aldrin proceeded with their pre-allotted chores. According to the Apollo 11 flight plan much of which Tom Reid knew by heart, their next three hours would be taken up with some still photography, as well as with aligning their guidance systems, loading the lunar module's computer with the necessary data for an immediate take-off if necessary, undertaking a simulated launch countdown and having something to eat. When all that was done, they would have been at it for a combined total of 10 hours since they had first begun preparing to undock from the command module for their flight to the lunar surface. Accordingly, NASA had scheduled a four-hour rest period to follow what it called their 'eat period'. After accounting for the length of time it would then take them to suit up for their Moon walk, Armstrong and Aldrin were not scheduled to emerge from their lunar module until about 4 pm.[7]

The only certainty for Tom Reid was that at Honeysuckle, the Moon would rise at 11.15 am. This meant that if the flight plan was followed, his team would have approximately two hours after acquiring the *Eagle*'s signal, to iron out any bugs before Armstrong and Aldrin began preparing for their Moon walk. After checking with Ian Grant that all was going well with the station's internal readiness testing, Tom ducked downstairs for a quick bite to eat. With the operations' shift yet to arrive and the maintenance shift now hard at work under Grant's supervision, the canteen was deserted, except for the catering staff who were just finishing the washing up. After helping himself to some bacon and eggs, Tom took a seat near the canteen's large picture window located at the opposite end of the building to the dish. On a clear day, this window provided a panoramic view back down the station's access road towards the valley below. But in this morning's pre-dawn gloom there wasn't much to see. Within a few minutes Tom expected that he would be able to pick out the fog lights of a convoy of cars as they raced up the hill.[8]

Although there was bunk-style accommodation at the station, just about everyone went home after their shifts. With around-the-clock activity throughout the operations building during an Apollo mission, Tom knew how hard it was to snatch any sleep; after a busy period on duty, his colleagues would be better rested after some time away from the place. He wouldn't have minded getting home to see Margaret and the kids either. But as station director he felt obliged to remain on call within the building during mission-critical periods.[9]

The station put a lot of effort into the logistics of ensuring that everyone arrived on time. To transport a shift of approximately thirty-five engineers and technicians backwards and forwards, Honeysuckle maintained a fleet of identical, current model Ford Falcon sedans, each of which bore sequential number plates. With four people allocated to each car, there were about nine vehicles all up. The person in each car who lived furthest away from Honeysuckle was normally designated as the driver, who would proceed to collect his three passengers with their order of pick-up dictated by how far away from the station they lived. Bryan Sullivan was due to be collected at around 6.15 am. As instructed, he had his porch light on so that the driver would know he was up and about. If someone did not have their light on, it meant that they had overslept and there would be a loud rap on their front door. Then they would be firmly told to come as they were and not worry about getting cleaned up until after they had arrived at the station.[10]

Although there was no particular reason for doing so, the station cars tended to form up into a convoy by the time they had reached Canberra's southern outskirts. After that they picked up speed as they roared by the entrance to Lanyon Station, thumped their way across the wooden bridge over the Murrumbidgee River and raced past the village of Tharwa. While everyone had been on time this morning and there was really no need to speed, it had become a NASA tradition to drive at a break-neck pace. Houston's white team leader, Gene Kranz, could remember that on his first day at Cape Canaveral back in 1960, he had been picked up by Gordon Cooper, one of NASA's original team of seven astronauts. 'My neck snapped back as Gordo floored the Chevy', Kranz recalled. 'I was thinking I had hitched a ride with a madman.' This craving for speed that began as a macho test pilot thing, had trickled down to most other levels of NASA's worldwide network including Honeysuckle. Local farmers along the way, like Frank Snow at Cuppacumbalong, had long since grown used to this. And knowing what would be at stake with Apollo 11 today, they felt a certain thrill as they watched this particular convoy charge on by into Canberra's hinterland.[11]

Racing up into the high country on a mid-winter's morning was not without its risks. There was always the possibility of colliding with a kangaroo or feral pig, sliding on black ice over the edge of the winding mountain access road and tumbling down into the valley below or being blocked by a drift of snow. Of greatest concern though, was heavy rain which could dislodge massive boulders above the road or wash it away altogether. One of the station's technicians, Hamish Lindsay, could recall a wash-away which had occurred immediately behind a convoy's lead car, stranding all the others on the Canberra side of the chasm. It took a bulldozer a whole day to carve out a temporary dirt track to reconnect Honeysuckle to Canberra. Had torrential rain been forecast for this morning, Tom Reid would have insisted that his operations shift remain overnight at the station as he had. But no rain of any consequence had been expected. As for the other risks, they were all avoidable. Despite his drivers' tendency to speed, Tom knew that they would all be keeping a special lookout for obstacles this morning. The usual topic of conversation on such drives was the previous night's TV. But today the discussion in Bryan Sullivan's car was firmly fixed on the mission's operations and procedures, instilling in everyone, including his driver, the need to arrive in one piece.[12]

Through the canteen's window, Tom was soon able to count off the lights of nine vehicles as one after another, they turned into the station's lower car park. Within minutes he felt a rush of freezing air as the thirty-five members of the operations shift followed each other into the building. After acknowledging those who stopped by the canteen to pick up a coffee and something to eat on the run, Tom went upstairs for the shift changeover which began at 7 am. Because everyone in the building was at a peak state of readiness after two years of intensive training, there wasn't much for Tom to do, other than to be ready to troubleshoot any unexpected problems. He simply stood back and observed. If the station had a peak hour, this was it as the maintenance and operations shifts mingled together. For about thirty minutes, there would be seventy people on the operations floor. And it wasn't hard to tell who was who. While the newcomers were fresh-faced, showered and shaved, those who'd been on duty overnight had dark lines under their eyes and jawlines covered in stubble. As the minty scent of fresh aftershave collided with the stale smell of lukewarm coffee and half-smoked cigarettes, each operations engineer and technician sat beside his maintenance counterpart for a full briefing on where, in relation to his particular console, computer or screen, the Apollo 11 mission was up to. This changeover lasted for about half an hour after which the dog-tired maintenance shift was driven off at high speed back to Canberra.[13]

As Tom Reid watched his team completing its station readiness testing, he wondered what surprises might be in store. Out of the corner of his eye Tom noticed that Bryan Sullivan had suddenly stepped back from his console and was holding his right upper arm out horizontally from his shoulder, with his forearm in a vertical position and his index finger pointing straight up from an otherwise balled fist. It was a sign to all to listen to Neil Armstrong speaking on Net 1, so Tom grabbed the nearest headset. It was just on 8.10 am.[14]

Within an hour or so of Armstrong and Aldrin completing their successful Moon landing, Gene Kranz and his white team had vacated their chairs at Mission Control. And in came Clifford Charlesworth, who together with his green team, were rostered to remain on duty until the end of Moon walk some twelve hours later. It was Charlesworth, who as NASA's senior flight director, had decreed this order of proceedings. Kranz later recalled that Charlesworth had summoned him to his office. 'I think it's time to decide on the Apollo 11 phase assignments,' Charlesworth had said. 'I think I should launch Apollo 11 and do the EVA [the Moon walk]. This leaves Glynn [Lunney] for the lunar ascent … and you with the landing. Is that okay with you?' As soon as Kranz had nodded his head, the meeting closed. It had lasted less than a minute. Who was this man, Charlesworth, who could summon and direct a Mission Control rockstar like Gene Kranz? He was, according to the black team's network controller, Richard Starchurski, a strange bird.

> The green team leader is a remarkably nondescript man
> with sandy hair and a medium build progressing toward the

> paunchy stage. His most distinguishing feature is an upper lip that he keeps twisting into an attitude of casual disdain. It gives fair warning of the type of encounter you're likely to have with the man. Sarcasm and abruptness are his constant companions.[15]

Among those assisting him on the green team were its CapCom, Bruce McCandless, a carrier-based fighter pilot who had joined NASA's astronaut intake in April 1966. Although he had not yet flown in space, McCandless was one of NASA's rising stars. And there was also the network controller, Ernie Randall, who Starchurski described as follows:

> Ernie Randall is a slender man with sandy hair and a crooked, boyish smile. If there is such a thing as a senior network controller, Ernie meets the requirements. He is a small-town boy from Ada, Oklahoma – an electrical engineer. He joined NASA in 1963, the same year that I transferred in. But he comes with extensive experience in mission control centre operations … In Houston he worked the network position on the first Gemini missions.[16]

If there was a senior flight control team at Mission Control during the Apollo 11 mission, then it was Charlesworth's green team. To them, Honeysuckle would be sending the *Eagle*'s downlink voice and telemetry data; and from them, Honeysuckle would be transmitting their voice and commands up to the lunar module. For the mission-critical Moon walk, Tom Reid and his team would be working hand in glove with Charlesworth, McCandless and Randall.

Although Apollo 11's 350-page flight plan, which had been signed off by NASA's director of flight operations Chris Kraft and its senior astronaut Deke Slayton, provided exhaustive details of every aspect of the mission including the Moon walk, and was for everyone including Tom Reid's team treated as something of a

Bible, it was not set in stone. During the Gemini missions which had preceded the Apollo program, Charlesworth and Kranz had come to realise the limits of the flight director's role. This had been especially so during the Gemini Spacewalks which were NASA's first extra vehicular activities (EVAs). 'During the EVAs, we could only listen to the crew and watch over the spacecraft systems,' Kranz later recalled. 'Only the commander's view from the cockpit afforded the perspective to make real "Go" or "No Go" decisions.' In the lead up to the Apollo 11 astronauts' Moon walk, the mother of all EVAs, Charlesworth had been quoted as saying: 'We want to stick to the flight plan. But you know how flight plans are – sometimes you have to change them.' Charlesworth also knew that during the final moments of the lunar module's descent, Kranz had yielded to Armstrong who had a cockpit window view. In this sense he understood as well as Kranz did, that the balance of power between Houston and Armstrong had shifted Armstrong's way. As things stood, the Moon walk was still scheduled to take place according to the flight plan: at about 4 pm. And there had been no change to this in the two hours that had elapsed since the lunar module had touched down. It was a fraction after 8.09 am when Bryan Sullivan, who had been listening to Net 1 from his Honeysuckle computer console, heard Armstrong speaking to Houston:

ARMSTRONG: Our recommendation at this point is planning an EVA with your concurrence, starting at eight o'clock this evening, Houston time. That is about three hours from now.

CAPCOM: Stand by.

ARMSTRONG: Well we will give you some time to think about that.

TRANQUILITY BASE, HOUSTON: We thought about it; we will support it. You're GO at that time.[17]

True to form, Charlesworth's decision had taken just 25 seconds, barely enough time for Tom Reid to grab a head set after seeing Bryan Sullivan's Net 1 arm and hand signal. As Tom well knew, bringing forward Armstrong's first step like this would soon trigger a mini tsunami of teletype messages from Houston, detailing the many alterations which would need to be made now that the flight plan's schedule had been compressed into a much shorter time frame. But until this flood of paper started to swamp Honeysuckle's communications area, Tom thought it best to let his station's readiness testing continue while he retreated to his office to think through the implications of an early Moon walk.[18]

A Prime Minister 'Blinded by Science'

After grabbing another mug of strong black coffee, Tom Reid settled in behind his desk. He lit a cigarette and gathered his thoughts. If the astronauts emerged in just 3 hours, at 11.15 am local time, Tom calculated Armstrong would begin climbing down the lunar module's ladder just as the Moon came in view of Honeysuckle's dish. There was every prospect that Honeysuckle would still be sorting out its TV downlink with the *Eagle* when Armstrong stepped onto the lunar surface. While Tom realised this would be a challenge for his station, he suspected that Clifford Charlesworth wasn't too concerned. No doubt the green team's flight controller was banking on Goldstone's multiple dishes being able to maintain links with both the lunar and command modules until mid-afternoon when their view period would end. These links would ensure continuity if Honeysuckle was still coming online at the critical moment. While Parkes's dish would miss out on much of the Moonwalk TV signal, Goldstone's big dish would be able to handle that too. The sceptic in Tom thought this would suit the Americans just fine. The glory of transmitting live TV of Armstrong's first step in prime time across the United States would go to Goldstone. Still, nothing was certain. Among other things, Tom thought, it might take the astronauts more than 3 hours to suit up before leaving the *Eagle*.[1]

By now the Honeysuckle team had completed their internal station readiness preparations and their computation and data flow of integrated sub-systems testing had begun. These so-called CADFISS tests were run by the Goddard Space Flight Centre in Maryland. Using pre-recorded test tapes, Honeysuckle's data flow paths for tracking, ranging, telemetry and command were thoroughly exercised and compared with known or predefined data values. Until Goddard was completely satisfied with the results, Honeysuckle would not be signed over to Mission Control. But even then, Houston would insist on running its own series of gruelling cross checks. According to Bryan Sullivan, these batteries of tests conducted firstly by Goddard and then by Houston, combined to produce 'one of the busiest and most activity intensive periods of Apollo tracking support'. Added to this was the anticipation of a flood of paperwork connected to the revised flight plan. With so much on their plate, Tom's team would not have relished the thought of the VIP visit which was about to descend on them – a visit which, as yet, they knew nothing about.[2]

From his office window Tom could see out over the operations building's main entrance and as far as the upper carpark beyond it. As he waited for his VIP to arrive, he was grateful that none of his team in the main operations area had this view. Only someone looking out through the cafeteria's picture window would have been able to see down over the entrance road. But the catering staff who had been banking on two orderly luncheon sittings in what had been expected to be a quiet period before a late afternoon Moon walk, were now scrambling to prepare for lunches on the run and were far too busy to take in the view.[3]

Just before 8.45 am, Tom noticed a four-car convoy proceeding in a stately fashion up the drive towards the main station entrance. In the lead was a magnificently polished Bentley which bore the number plate C1 and flew the Australian flag from a small pole attached to its bonnet. After moving smartly down the station's front steps, Tom offered a firm handshake to the Prime Minister of Australia, The Right Honourable John Grey Gorton, PC, MP. Travelling with him were his press secretary, Tony Eggleton, and his private secretary, Ainsley Gotto. The next car carried a security detail which had been beefed up following Harold Holt's drowning eighteen months earlier. Emerging from the car behind that were Alan Cooley and Lloyd Bott who were respectively the secretary and deputy-secretary of the Commonwealth Department of Supply, together with Willson H. Hunter who was NASA's senior scientific representative in Australia. Alighting from the fourth vehicle were an ABC TV cameraman, a sound recordist and a reporter. In terms of Tom Reid's superiors in Australia, it was the most formidable group imaginable – all to be filmed by ABC TV.[4]

Tom's wife, Margaret, had known John Gorton since at least 1961 when as Minister for the Navy, he had helped to convince her to stand for the hard-luck Adelaide seat of Bonython. In recent years, as Margaret had continued her strong association with the Liberal Party in Canberra, she and Tom had been entertained by Gorton and his American-born wife, Bettina, at their Canberra home in Narrabundah and more recently at The Lodge. Margaret was also a good friend of the Gortons' daughter, Joanna. But this visit was official. And it had been organised the afternoon before following a phone call to Tom from his superior, Lloyd Bott. Sensing Tom's unease which Bott shared, the deputy-secretary had tried to sound reassuring. 'The Prime Minister wants to highlight the key role your tracking station will be playing tomorrow,' Bott said, 'so he will be bringing a TV crew with him.' Far from being placated, Tom wondered what sort of equipment this crew might be carrying and how it might affect the station's sensitive instruments. He knew his team would be unimpressed by such an awkwardly timed prime ministerial appearance, less than three hours before Honeysuckle acquired the lunar module's signal. He felt he had no choice but to agree to Bott's request. While his team's focus was wholly on the mission, Tom's responsibilities as station director necessarily involved maintaining cordial relations with the highest levels of the Australian and American governments. If the Prime Minister wanted to undertake a station visit on the morning of the Moon landing he'd do his best to ensure its success. What Tom hadn't been able to factor in was Armstrong's decision to bring forward his Moon walk by five hours.[5]

To keep disruption to a minimum, Tom had told no one at the station about Gorton's visit. And he tipped off the security staff only moments before the Prime Minister's motorcade had first come into view. If his team knew about it, he reasoned, they'd start thinking, talking and even possibly complaining about it. Advance notice of this VIP visit was an unnecessary and avoidable distraction. Indeed, Tom had gone so far as to refrain from getting one of the station cars to stop by his house for his suit, concerned that this might tip off the driver that something was up. While everyone else in the official party including the people from the ABC wore a coat and tie, Tom made do with his smart casual rig: slacks, an open necked shirt, a cardigan and a polka-dot cravat. To minimise the disruption, the party that entered the station's main operations area was culled down to the Prime Minister, Tony Eggleton, Alan Cooley, the ABC journalist and his cameraman. Because his equipment might affect the station's more sensitive machines, the ABC's sound recordist was left to cool his heels in reception along with Ainsley Gotto and the remainder of the official party.[6]

The surprise of this visit was so great that for a while many of the engineers and technicians did not realise the Prime Minister was in their midst. Intent on watching their various screens and listening to whatever was coming through their headsets, people like Bryan Sullivan were astonished to look up and see John Gorton standing right beside them. Normally mild mannered, Sullivan's initial reaction was: 'What the hell is he doing here?' As the Prime Minister walked by each of the operations areas, Tom remained by his side attempting to explain as best he could the ins and outs of Unified S-bands, uplinks, downlinks, telemetry, demodulators, Univac computers, carriers and sub-carriers. After using his hands to help illustrate the points he was trying to make, Tom would move his head back just a fraction as he subtly attempted to assess how much of what he was saying the Prime Minister was able to take in. Like most visitors to the operations area before him, John Gorton was drawn to the servo console where Paul Mullen explained how he could control the dish's speed and angle by rolling the servo ball with the palm of his hand. And then like most other visitors before him, the Prime Minister took his turn at rolling as he looked out the window to see how the dish responded.[7]

Another area that seemed to hold the Prime Minister's attention was the operations control room, the nerve centre for the whole station, where Ian Grant sat in the Ops 1 seat with John Saxon beside him. While all the other areas were open plan, the operations console was set up in this special room which had an internal glass window through which whoever was in charge could see what was going on elsewhere. But even though it was a confined space, John Saxon had a similar experience to Bryan Sullivan. Saxon was so intent on the console in front of him that he did not at first see the Prime Minister standing behind him. And when he did, it was lucky that he didn't verbalise his thoughts. Known throughout the station for his effervescent, wisecracking, irreverent sense of humour, any comment Saxon may have made would not have gone down well with the official party.[8]

Like any other politician, John Gorton was fascinated by television so he spent some time talking to Ed von Renouard, the technician who operated the station's slow scan converter which would render live TV signals from the lunar module into a form suitable for commercial television. Born in Germany, Ed was a radio technician who had emigrated to Australia to avoid the bleakness of post-war Europe. With the goal of wanting to go somewhere 'you couldn't buy a ticket to', he ended up in Antarctica, recording signals being sent back to Earth by weather balloons. After returning to mainland Australia, Ed had studied for his Television Technician's Certificate at the South Australian School of Mines and Industries and then worked for the ABC in Adelaide, before moving to a job in telemetry at the Island Lagoon tracking station near Woomera. From there he moved to Honeysuckle Creek. With his background in television, he was immediately sent over to Goldstone to be trained up on the newfangled slow scan converter which had only recently been installed there. After returning to Honeysuckle, Ed had been involved in the Apollo 8, 9 and 10 missions. With a noticeable German accent and a modest manner, he wasn't the easiest person to talk to. But the Prime Minister was fascinated by his story and listened intently as among other things, Ed explained the significance of a small metal switch. Located on his console just above head height, this switch was identified by two small Dymo labels which were stuck immediately above and below it. Silver in colour, such switches could in those days be purchased at a hardware store for just a few cents. But as Ed explained to Gorton, this particular switch was attached to an apparatus designed to flip an upside-down live TV image from the Moon, right side up, so as to make it intelligible to millions of viewers in their living rooms. The Prime Minister also marvelled at the tiny TV monitor screens built into Ed's console. From these, Ed explained, he'd be able to see a fraction of a second before the rest of the world, Neil Armstrong stepping onto the Moon.[9]

After moving outside to a point where the ABC's TV camera would be able to capture the moving dish behind him, the Prime Minister prepared to read a statement and take questions. When it came to TV stories about tracking stations, television journalists believed that their viewing audiences would be bamboozled by descriptions of highly technical equipment so they demanded colour and movement. Shortly before the Apollo 11 Mission had begun, the legendary CBS Evening News anchor Walter Cronkite, had spent a few days at Goldstone where the intricacies of space tracking were explained to him and much film was shot inside the station's operations area. But what CBS put to air were poetic long shots of the Goldstone dish moving at speed against a backdrop of its arid landscape, with Cronkite waxing lyrical about the beauty of the surrounding desert, interspersed with musings about the meaning of Apollo 11 for humanity. Any details of how Goldstone actually worked had been ruthlessly edited out.[10]

With its lights flashing, the Honeysuckle dish was angled down behind the Prime Minister who read a prepared statement, praising the contribution Australia was making to the Apollo 11 mission through its tracking station network. The subtext of this was an acknowledgement of Australia's alliance with the United States. Unlike his predecessor Harold Holt, John Gorton had not gone 'all the way with LBJ' and as a consequence, his relationship with President Johnson had been frosty. With Johnson's successor, Richard Nixon, Gorton had been keen to make up lost ground. During his visit to the White House in June 1969, he had addressed President Nixon: 'Sir,' the Prime Minister said, 'We will go Waltzing Matilda with you.' With a Federal election due before the end of 1969, Gorton had remained keen to emphasise the importance of this special relationship, a stance which was still popular with that segment of the voting public the Liberal Party relied on. While admitting that he had not woken up early enough to hear a live broadcast of the lunar module's Moon landing, the Prime Minister was nevertheless fulsome in his praise: 'The United States has achieved a great and peaceful triumph for mankind generally,' he said. Then it was time for questions. To provide context for the Prime Minister's own technical prowess, the reporter noted that during World War II Gorton had flown Hurricane fighters at 350 miles per hour. Gently correcting him, Gorton pointed out that a Hurricane's normal operating speed was 180 miles per hour. 'Yes, they could be flown at 350 mph,' he conceded, 'but only for about thirty seconds before their engines blew up'. After mentioning that his sound recordist had been excluded from the tour of the operations area, the ABC reporter asked what the Prime Minister had discussed with the station staff. To this Gorton responded:

> I asked what all the little wiggly green lines were and what all the little noises coming out of the bits and pieces of equipment were. I got answers I couldn't thoroughly understand. I was blinded by science.

After getting the obligatory photo in front of the dish with Alan Cooley, Lloyd Bott and Tom Reid beside him, the Prime Minister wished Reid well and took his leave. Gorton's station tour and press conference had lasted just on an hour.[11]

By the time Tom Reid was able to settle back into the station's rhythm of readiness testing, it was almost 10 am. And he was relieved to see that after their brief chats with the Prime Minister, the station's engineers and technicians had quickly refocused on their CADFISS tests. The latter stages of these tests with Goddard were so intensive that Bryan Sullivan likened them to 'the demands made on air traffic controllers on a Friday afternoon of a long weekend at the start of the school holidays'. Tom's team also appeared unfazed by the blizzard of amendments to the Apollo 11 flight plan which over the next hour landed on their consoles. Multiple copies of these had been run off from the original teletype messages that had flooded the station's communications area following NASA's decision to bring forward the Moon walk. Each copy then had to be cut and pasted onto replacement pages, to be distributed to every one of the station's manuals right down the line. At around 10.30 am as the CADFISS tests wrapped up, Tom Reid's other deputy, Mike Dinn, replaced Ian Grant in the Ops 1 seat, bringing an end to Grant's gruelling 12-hour shift.[12]

With a strong West-Yorkshire accent, Mike Dinn hailed from Bradford. After obtaining an electrical engineering degree, he had specialised in aircraft electronics before moving across into space tracking. Solidly built, Mike had the physique and apparent strength, especially in his wrists and forearms, of someone who might easily have worked in one of Bradford's coal mines. A deliberately spoken man who carefully considered each word he uttered, Dinn had a complete grasp of Honeysuckle's key systems and all who operated them. He was renowned, especially on the organisational side, for coming up with innovative solutions. In the lead up to Apollo 11, Honeysuckle had been bombarded with media requests to film the station. Rather than have the station's work constantly interrupted by film crews, Mike had proposed that Honeysuckle hold a special media day where the dish could be driven at speed for the TV cameras. Tom Reid adopted this proposal with alacrity and the media happily filmed away as the dish was put through its paces. Although the dish was never driven this fast during the real time tracking of Moon missions, the media lapped it up all the same. Even so this measured approach to media management had not been enough for the Prime Minister. For John Gorton, nothing less than TV images of him inspecting the station on the day of the Moon walk, and having the dish moved out of its Moon tracking alignment for a photo opportunity, had been sufficient.[13]

Just on 10.45 am Goddard was heard on Net 2: 'HONEYSUCKLE, NETWORK. That completes interface testing, your site is GO. Standby for Houston.' Then, over Honeysuckle's paging system, Mike Dinn announced: 'All positions, Honeysuckle is at H–30.' This meant that in just thirty minutes, the Moon would have risen sufficiently for Honeysuckle to be able to lock onto the lunar module. For the next twenty minutes, Mission Control conducted final checks: voice circuits all the way from Houston to Honeysuckle's transmitters were tested; the formats from Honeysuckle's telemetry computer were verified; and bursts of digital commands were fired through Honeysuckle's command computer. At the servo console, meanwhile, Paul Mullen had received the latest antenna-pointing data tape derived from Goldstone's most recent signals. And at H–10, as Bryan Sullivan put it: '[Honeysuckle's dish] rolled across the sky and pointed down to focus on the exact spot on the eastern horizon where the Moon would rise.' Meanwhile a red light down near the station's main entrance gate had begun to flash and all vehicles had to stop, a large sign directing drivers to turn off their engines lest their electronic ignitions interfere with the station's ultra-sensitive receivers. Mike Dinn waited for Houston's call on Net 2.

HONEYSUCKLE, NETWORK. Station status?

NETWORK, HONEYSUCKLE. Our status is GREEN.

All the while, Tom Reid had been quietly observing how well Honeysuckle had handled Houston's last-minute tests. The Prime Minister's visit had not fazed Tom's team. And as the final seconds ticked away he felt intensely proud of them. Just on 11.15 am, the Moon rose beside Dead Man's Hill.[14]

God Damn It: We Were Ready!

For at least 21 000 years before the arrival of the first Europeans in the 1830s, members of the Namadgi tribe had stood on the future site of the Honeysuckle Creek tracking station and stared in wonder at the Moon. Like almost everyone else who had ever lived from prehistoric times up until the late 1950s, they never imagined that one day a human being would set foot on it. That moment was almost at hand and Honeysuckle's dish was about to come online to support Neil Armstrong's first step.[1]

Even though history had not been Tom Reid's best subject at school, he understood that he and his team would be playing a significant part in the climax to the greatest voyage of exploration humans had ever attempted. He had appointed his ranging and timing technician, Hamish Lindsay, to chronicle this moment. Raised in Tasmania, Lindsay was an accomplished commercial photographer who had been allowed to convert one of Honeysuckle's downstairs bedrooms into a dark room. With Tom's blessing, he had followed the Prime Minister around that morning, reeling off film of John Gorton's visit. For the rest of the day Lindsay had a roving commission to photograph and write about what unfolded, including the moment Honeysuckle's dish acquired the lunar module's signal.

> At 11.15 am local time, the Moon rose above the gum-tree clad horizon beside Dead Man's Hill until the Honeysuckle Creek receivers promptly locked on to the *Eagle's* signals. They flooded through the station's equipment, kicking meter pointers up scales, rolling figures around readouts. Anxious eyes rapidly scanned over the panels, watching until they all settled nicely down to normal readings. Everything was working perfectly.[2]

Honeysuckle's 85-foot dish was locked onto the lunar module's tiny 26-inch antenna. Downlink voice and telemetry signals streamed into the station which after being processed, were sent on to Houston via Goddard. Because the lunar module was in a fixed and stable position on the Moon's surface, Tom Reid was relieved to see that there were none of the intermittent drop-outs that had plagued much of its descent phase. When in mutual view of the Moon as they were now, Honeysuckle and Goldstone were technically capable of simultaneously maintaining uplinks and downlinks with the lunar module. While they could each receive downlinks at the same time without causing any problems, their uplinks if transmitted together were more than the spacecraft's tiny on-board computers could handle. As a result, vital voice and command signals from Mission Control to Armstrong and Aldrin would become scrambled. According to the black team's network controller, Richard Starchurski, 'Two carriers on the same frequency turn the whole thing to worms.' One of the most rigorously enforced protocols throughout the worldwide tracking network was a precisely choreographed series of steps, described by Starchurski as 'a duet', to transfer the uplink signal from one station to the next. As Goldstone was generating a good uplink signal to the lunar module and would continue to have the Moon in view for some hours yet, Mission Control decided to leave this arrangement in place for the time being.[3]

When Houston had agreed to the astronauts' request for an early Moon walk, Armstrong predicted that he would leave the lunar module in about three hours. But that time which coincided with Honeysuckle acquiring the lunar module's downlink signal had now come and gone. There was still no sign of them as they continued the gruelling but critically important process of suiting up. An astronaut who emerged from the lunar module in a malfunctioning spacesuit risked a horrible death. As Bryan Sullivan described:

> Your brain would be immediately starved of oxygen and you would lose consciousness. Your lungs would then collapse, the air in your heart would bubble, and your blood vessels would rupture. You would die within a minute or so.[4]

Armstrong and Aldrin were painstakingly careful with their preparations. After entering their custom-made spacesuits from the rear, the first thing they did was to put on their overshoes, the coarse-treaded soles of which would give them a firm footing on the Moon's surface. Next, they strapped on their massive backpacks. Despite weighing very little in space, these were bulky and they were vital because they contained the equipment that generated Earth-like conditions for their wearers. Linking them to the front of the astronauts' spacesuits were a number of hoses. One of these carried oxygen, while another carried cooled water that circulated through tiny tubes woven into an undergarment. Each hose had to be locked into its own special metal portal. For added safety, the hoses were double locked in a procedure which Armstrong checked off with Houston via the tracking network's voice downlinks: 'Locks are checked, blue locks are checked. Lock-locks, red locks, purge locks are checked.'[5]

Next came the astronauts' transparent bubble-like helmets over which were fitted outer layers of protection, including tinted visors to reduce the Sun's glare. After that Armstrong and Aldrin checked the fronts of their suits where, at chest height, there were small boxes that controlled their radios and displayed data readouts from their backpacks. As each task was completed, the astronauts were a little bit closer to turning themselves into fully self-contained mobile spacecraft. But they were spacecraft that needed to keep in touch with each other and with Mission Control. More than ever, Houston needed reliable telemetry, especially relating to their heart and respiration rates, to their temperatures, and to the oxygen and pressure levels inside their suits. The connections that had allowed their voices and personal telemetry to flow from the lunar module down to Honeysuckle, now had to be transferred across to these suits. The mode, order and sequencing of each of these transfers had been carefully set out in the flight plan. But as Kevin Gallegos attempted to keep up with the astronauts from his demodulator console, they departed from their script. John Saxon, who at that time was in the Ops 2 seat, later recalled:

> The checks on the portable life support systems at this point were in a totally different sequence to what we were expecting – every time they changed modes we had to make major reconfigurations on the ground – we were really, really busy trying to keep up with the astronauts doing their own thing. The busiest man without question was Kevin Gallegos... because all these modes affected how he routed the signals through the station and he had to literally second guess what the astronauts were doing.

Tom Reid's faith in this knockabout former petty officer with the razor-sharp mind had not been misplaced. Although Gallegos later admitted to breaking out into 'a cold sweat' while he 'whipped things around', it only took him ten seconds to get the new voice and telemetry links running smoothly once the astronauts had connected themselves up.[6]

It was after 12.30 pm. Well over an hour had passed since Honeysuckle had locked onto the lunar module's downlink. But there was still no sign that the astronauts would emerge anytime soon. In another thirty-five minutes, the Moon would have risen sufficiently over Parkes for its 210-foot dish to be able to pick up this downlink signal too. Although Honeysuckle and Parkes were at roughly the same longitude, Honeysuckle's dish could be angled down to zero degrees: to the horizon. But the Parkes dish could only be angled down to about 30 degrees above the horizon. NASA's rule of thumb was that each 15 degrees' extra elevation equalled one extra hour until Moon rise. This is why Honeysuckle's dish was able to lock on to the *Eagle*'s signal almost two hours earlier. But as soon as the Parkes dish did lock on, its voice and telemetry signals would start flooding into Honeysuckle for processing. Maintaining close contact between Honeysuckle and Parkes was a vitally important job that Tom Reid had assigned to his deputy director, Mike Dinn. Since the completion of the station's CADFISS tests at 10.30 am, Dinn had been in the Ops 1 seat which he vacated for Tom to concentrate on Parkes.[7]

Honeysuckle's operations control room was dominated by two adjoining consoles which combined to link all the station's functions together. The one on the left was the domain of the senior managerial person on duty, designated Ops 1, whose responsibilities included all operations relating to tracking, signal reception and signal processing. And the one on the right was operated by a senior staff member, referred to as Ops 2, who kept an eye on the data processing of all telemetry, command and voice communications entering and leaving the station. Built into each console was a range of lights and gauges, including volt meters which showed the strength of the signal in each of the dish's receivers. Others gave read outs on the dish's transmitters, indicated the status of data flow paths or displayed the station's overall configuration. These consoles also provided complete access to all the station's intercom loops as well as uplink and downlink voice communications. Anyone within the station could be contacted from these consoles, as could Clifford Charlesworth and his green team in Houston, and in an emergency, the astronauts themselves. Finally, there was a screen to monitor TV images coming in from the Moon. Ironically, Honeysuckle was one of the few places in the Canberra region that was unable to receive a commercial TV transmission.[8]

Sitting in the Ops 2 seat beside Tom Reid was John Saxon. Born and educated in Britain, Saxon was an engineer who had been involved in the development of inertial navigation systems and later in the testing of air-to-ground missiles at Woomera. For the Apollo 8, 9 and 10 missions, he had acted as Honeysuckle's Ops 2. Talkative, opinionated and with a cheeky sense of humour that could get him into trouble at times, Saxon wasn't a typical tracking station Ops 2 in the Goddard/Houston mould where crisp, humourless personalities were more prized. He had proved that he could do the Ops 2 job and do it well and that's all that mattered to Tom, whose Glaswegian sense of humour NASA couldn't understand either.[9]

As Reid and Saxon listened in to the astronauts doing their final voice checks, intermittent interference crackled through their headsets. Their first thought was that something at Honeysuckle was malfunctioning. But they were relieved to figure out the cause of the problem: the astronauts' personal antennas which protruded above their backpacks, were scraping against the ceiling of the *Eagle*'s cabin. By this time Armstrong and Aldrin were venting the cabin's atmosphere to reduce the pressure on its exit hatch. But they struggled for a few moments to open it – a metal door so thin that neither of them wanted to force it lest they wreck it. Then Aldrin peeled back one of its corners to break the seal. When it yielded, what was left of the cabin's atmosphere escaped outside and immediately froze into tiny flakes of ice. Armstrong then manoeuvred his body outside onto the platform known as 'the porch' which was connected to a ladder extending down one of the *Eagle*'s landing struts.[10]

Around the world over 600 million people were glued to TV screens. Among them was John Gorton. Having just given his most senior ministers a briefing on his visit to Honeysuckle that morning, the Prime Minister and his cabinet had adjourned to an anteroom in his Parliament House office where a TV had been set up in the corner. Margaret Reid, meanwhile, had left her Canberra City legal office early and had driven home to Hughes. There she joined her daughters, Marg and Danae who, like their classmates, had been given the afternoon off school. Having settled into their lounge room chairs they waited for the TV broadcast to begin. The boys, Tommy and Nick, remained at school; Nick in his school's assembly hall where he and his friends sat waiting too. Although they didn't discuss their dad's work much at school and didn't understand a lot of it, Tommy, Marg, Nick and Danae were proud that the station he led was supporting Apollo 11 today.[11]

As Neil Armstrong crawled backwards across the *Eagle*'s porch towards the top rung of the ladder, Tom Reid realised that the TV broadcast from the Moon would begin some minutes before Parkes's big dish could lock onto the lunar module's downlink signal, that is some minutes before the Moon rose over the Parkes dish at its lowest angle of 30 degrees above the horizon. There was a chance that Parkes's off-axis receiver might get a TV signal before that. But Tom knew that this signal would be unstable, possibly jerky, and prone to drop in and out. It would not be up to Mission Control's broadcast standard. Sydney Video's Charlie Goodman would choose Honeysuckle's more stable signal to send on to Houston, at least until Parkes's main dish could get a signal. This meant that for Neil Armstrong's first step, Honeysuckle's 85-foot dish would be the only available TV back-up for Goldstone's 210-foot dish, which because it still had the Moon well in view, had just been earmarked by Houston as the prime station for live TV. Tom switched on his station-wide intercom and momentarily suppressing a feeling of rapidly rising excitement, he barked 'Battle Short!'[12]

Although Tom Reid's primary focus in these final moments was on whether the crucial downlink voice and telemetry signals would continue to work as the astronauts began to walk away from the lunar module, he also wondered how its remote TV camera would perform. Only a couple of weeks had passed since NASA had publicly committed to televising Armstrong's first step live and there had been no opportunity to fully test and rehearse the key systems required to bring this moment into people's living rooms.

There would be no clapper boards, countdowns or hand signals to indicate when the lunar module's TV camera would begin filming; and no one at NASA had any idea what its pictures might look like. The challenges of the split-second decision making that lay just ahead were enormous. Regardless of how these decisions might be made elsewhere, and Goldstone faced the same challenges, Tom was confident that at Honeysuckle, his TV technician, Ed von Renouard, would rise to the occasion. Satisfied, he pressed his intercom button and summoned Hamish Lindsay. 'Just before Armstrong took his first step on the lunar surface,' Lindsay later recalled, 'Tom Reid sent me out to record the moment. It was wet and cold and mid-winter – we were suffering sleet showers at the time.' Moving away from the station campus area and into the bush beyond, Lindsay snapped off an iconic shot of the dish, framed in the background by a bleak sky and misty mountain tops and in the foreground by gum trees and boulders. Angled up, over and away from where Lindsay was standing, the dish presented an almost poetic mix of gentle curves and sharp angles. It was immensely strong yet delicate. And quite apart from its flashing lights, its overall stance signalled that it was ready to receive live TV from the Moon.[13]

Before taking his first step down the 8-foot ladder, Armstrong pulled on a D-ring which was attached to a lanyard. This activated the *Eagle's* MESA stowage bay which swung out and down to reveal the small TV camera inside it. Due to earlier overheating, the TV circuit breaker had been removed. At first, nothing happened. Mission Control called up the astronauts on Net 1.

> HOUSTON: Neil, this is Houston. You're loud and clear. Break. Break. Buzz, this is Houston. Radio check and verify TV circuit breaker in.

> ALDRIN: Roger. Roger TV circuit breakers in and read you loud and clear.

Almost immediately, an image appeared on Ed von Renouard's slow scan converter. For a split second what Ed saw confused him.

> It was an indecipherable puzzle of stark blocks of black at the bottom and grey at the top, bisected by a bright diagonal streak. I realised that the sky should be at the top, and on the Moon the sky is black, so I reached out and flicked the switch and all of a sudden it all made sense, and presently Armstrong's leg came down.

As had been planned so as to save room in the stowage bay, the lunar module's TV camera began automatically filming from an upside-down position. To compensate for this, Ed had remembered to engage the toggle switch on his scan converter which immediately flipped the TV picture the right way up. That done, Honeysuckle began transmitting broadcast quality pictures to Houston. Tom Reid was well pleased with the TV footage which filled the little screen on the monitor in front of him, showing Neil Armstrong slowly making his way down the *Eagle*'s ladder.[14]

At Goldstone, however, confusion had set in immediately after Armstrong pulled the D-ring. The station's prime shift had gone off duty. It was one of the technicians on the back-up shift who had been left in charge of Goldstone's scan converter. Stan Lebar later recalled:

> I was listening to the communications traffic and heard someone say, 'Make certain the reverse switch is in the reverse position'. A couple of minutes later a second voice made the same announcement and a bit later another voice repeated the message. After that it would be a miracle if the switch was in the correct position as I could almost visualise someone throwing the switch each time the message was sent. I suspect by then the operator was so hyper that he didn't know what position the switch was in.[15]

By now thoroughly confused, the Goldstone technician increased the contrast on the scan converter's output, dragging most of the picture into the black and making it very high contrast. Still struggling, his next mistake was to adjust the focus, the result being that the picture was not as sharp. And a little while after that, he tried another setting, turning the picture to negative. This had the effect of compressing the shadow areas into white. With each mistake, the technician was compounding his incorrect settings.[16]

The responsibility for choosing which TV images to put to air rested on the shoulders of Mission Control's Ed Tarkington, known by his call sign 'Houston TV'. Arrayed in front of Tarkington, in a room behind the main Mission Control room, were a number of monitors showing the TV images coming in from the tracking stations that had the lunar module in view. Generally speaking the bigger the dish, the better the TV image. Tarkington could not at first understand why the TV pictures he was seeing on his monitor from Honeysuckle, that had already been preselected by Charlie Goodman in Sydney, were so much better than the ones he was seeing from Goldstone. How was it that Tom Reid's team, where there was not an American accent to be heard, was outperforming NASA's prime station in California with its much larger dish? Putting his faith in the bigger dish, Tarkington had at first chosen Goldstone's TV images to put to air. As he continued watching, it was impossible to make out what Armstrong was up to. Knowing that the director of flight operations, Chris Kraft, had thrown his considerable prestige behind the last-minute decision to televise Armstrong's first step live, Tarkington instinctively understood that Kraft would be as unhappy as he was with Goldstone's indecipherable pictures going to air. Still he gave Goldstone one last chance. And seventeen seconds after the Lunar Module's camera had begun filming, Tarkington contacted Goldstone on Net 2:

HOUSTON TV: Goldstone video, Houston TV.

GOLDSTONE: Goldstone video, go ahead.

HOUSTON TV: Can you confirm that your reverse switch is in the proper position for the camera being upside down?

GOLDSTONE: Stand by, we will go to the reverse position…We are in reverse.

HOUSTON TV: Roger, thank you.[17]

Thirty-one seconds had now passed since the beginning of filming and Goldstone was still having trouble. Meanwhile, oblivious to all this drama at Mission Control, Armstrong had continued to make cautious but steady progress down the ladder. The *Eagle*'s landing struts were designed to telescope, cushioning any landing impact. But having landed gently, Armstrong found that there was a larger gap between the ladder's bottom rung and the Moon's surface than he had planned for. He dropped off the bottom rung onto the strut's footpad and then immediately jumped back up again to see whether he could overcome this gap. 'It takes a pretty good little jump,' he said. After that, he stepped back down onto the footpad.[18]

At Honeysuckle, staff had come from all over to stand silently behind Ed von Renouard's scan converter, their eyes glued to his TV monitor. Secretaries, cooks, gardeners and many, many others stood stock-still and silent, like a column of terracotta warriors. But there were too many of them for the confined space. And one minute and thirty-six seconds after filming had begun, there was an announcement over the station's intercom: 'Will all personnel move away from the scan converter please.' A further nine seconds after that, Ed Tarkington was heard on Net 2:

HOUSTON TV: All stations, we have just switched video to Honeysuckle.[19]

Having waited some more for the team at Goldstone to come good, Tarkington had finally given up on them. It was thanks to Honeysuckle that, for the first time, 600 million people could clearly make out Neil Armstrong. Among them were a lunchtime crowd of hundreds rugged up in their winter coats, who milled around the David Jones Department Store in central Sydney, to watch TVs set up in its display windows fronting Elizabeth and Market Streets; and the tens of thousands who crammed into New York's Central Park, on what over there was a sultry summer's evening, to view a bank of massive TV monitors specially erected by the CBS Network. This worldwide TV audience had just heard Armstrong say: 'I'm at the foot of the ladder. The LM footpads are only depressed in the surface about one or two inches, although the surface appears to be very, very, fine grained…' It had been at this point in Armstrong's voice transmission, one minute and forty-five seconds after filming had begun, that Honeysuckle's TV signal had first been seen by the world. For another twenty-five seconds, Armstrong continued to hold onto the lunar module's strut. And with one of his boots still on the *Eagle*'s footpad, he tested the lunar dust with the tip of his other boot. Then he said: 'As you get close to it, it's almost like a powder. The ground mass is very fine…I'm going to step off the LM now…' What followed was a short pause in Armstrong's voice transmission during which Honeysuckle's TV signal showed him letting go of the strut and stepping backwards to plant his left foot in the lunar dust. It was 12.56 pm at Honeysuckle when Armstrong said on Net 1:

> That's one small step for (a) man, one giant leap for mankind.[20]

Tom Reid and his Honeysuckle team had nailed – they had absolutely nailed – live television of this supreme moment: the culmination of more than a decade's work by almost half a million people, when a human being had, for the first time in history, set foot on a celestial body. In New York's Central Park, the massive crowd which was bitterly divided over the increasingly bloody Vietnam War, united in this moment of triumph, their faces lit up by various states of ecstasy, joy and wonder. For those who were privy to the telemetry streaming into Honeysuckle from Armstrong's spacesuit, his heart was beating 112 times per minute, compared to Aldrin's 81. While the astronauts were clearly elated, Goldstone's lead engineer, Bill Wood, was despondent:

> I saw the network TV here – we were picking up the commercial television out of Los Angeles and when we saw the switch from Goldstone to Honeysuckle there was a pronounced improvement in the video quality. 'Hey, look at the picture from Honeysuckle!' and I thought 'Good Lord there's something wrong with our system – they are getting it much better than we are.'[21]

Even so, Houston would not give up on Goldstone. Just over three minutes later, Ed Tarkington announced 'All stations, Houston TV. We have just switched to Goldstone video.' Goldstone, however, was still in a mess so Tarkington was back online in less than a minute:

> HOUSTON TV: Honeysuckle, Houston TV. We have just switched back to you again.

Following his prompt for Goldstone to 'check your polarity switch, please,' Tarkington crossed back over to Goldstone once more just over a minute later, in the hope that this station's much bigger dish might finally be transmitting the higher quality TV images it was capable of. But things were still far from perfect. Many years later Stan Lebar recalled:

> When Armstrong started his descent from the bottom rung [of the ladder] most of us couldn't help but laugh as we all remembered the conversations and comments on the

> probability when it was first suggested that we mount the camera upside down and rely on a reversing switch at the tracking stations.

The reversing switch, however, had presented no problems for Ed von Renouard. And Honeysuckle's live TV feed continued to go to air worldwide, until the main signal from Parkes finally came online, 8 minutes and 53 seconds after the lunar module's camera had begun recording and over 6 minutes after Armstrong had taken his first step. From then until the end of the astronauts' Moon walk, Tarkington elected to use Parkes's stronger signal generated by its larger dish for the worldwide TV broadcast.[22]

The only voice transmissions which the public had heard on TV were those on Net 1 between Mission Control and the astronauts. NASA had not broadcast the Net 2 exchanges between Houston TV, Goldstone and Honeysuckle.

Like my school friends and I had noticed in the Sydney Grammar science auditorium, the worldwide TV audience would have seen a marked improvement in picture quality immediately after the switch to Honeysuckle had been made. The reason for this improvement was known to only a handful of people, such as Houston's mission controllers, Stan Lebar, and Bill Wood at Goldstone, who had been able to listen to Ed Tarkington on Net 2 and watch Los Angeles television at the same time. Even fewer of Tom Reid's team knew.

After Honeysuckle had settled more comfortably into its TV transmission, Tom had called up his section heads, allowing them to release one person at a time 'to go and look at the TV [monitor] if they want ... just for a few minutes and back'. Because Honeysuckle was beyond the range of Canberra's television channels, this was all most of them saw live. Only a few of them had been able to listen to Ed Tarkington on Net 2. Accordingly, most people at Honeysuckle had no idea that they had been the ones to transmit the live TV of Armstrong's first step. Nor did Tom tell his family. Only rarely did he talk about his work at home and he wasn't one to boast. When he had told his mother that one day a man would walk on the Moon, Mary Reid had dismissed him with a curt 'Och, Tommy!' Now as she watched Armstrong on TV, she shook her head in disbelief. 'Imagine that!' she said. But Mary never imagined that it was her son who had brought Armstrong into her Launceston living room. It wasn't until a function at Canberra's Lakeside Hotel in 1989, to celebrate the twentieth anniversary of the Moon landing, that Tom finally let his hair down during a speech in which he reminisced about how Honeysuckle had brought Armstrong's first step to the largest worldwide TV audience in history. 'It hadn't been planned that way,' he said of his team's achievement. 'But that's the way it was. And God damn it, we were ready!'[23]

Epilogue

With the benefit of almost fifty years' hindsight there can be no doubt the high point of manned space exploration was when Neil Armstrong took that first step. As Gene Kranz wrote in the year 2000: 'It was worth every sacrifice.' Although the Americans never publicly acknowledged how Honeysuckle rather than Goldstone had brought that moment to the world, Clifford Charlesworth came close. In a congratulatory telex to Tom Reid on 24 July 1969, the day Armstrong, Aldrin and Collins had safely returned to Earth, the flight director said:

> The flight operations personnel fully appreciate the tremendous effort required to achieve the readiness posture which made possible the greatest television spectacular of all time.[1]

In terms of Tom's future career, the most important praise came from the permanent head of the Department of Supply, Alan Cooley, who had accompanied Prime Minister Gorton on his station tour. After noting that Honeysuckle's performance during Apollo 11 had earned the warmest commendation of NASA's mission controllers, Cooley told Reid:

> Apollo 11 was the mission for which the station was created and I have watched...[its] development and improvement with great interest. All the hard work, training and effort was rewarded during Apollo 11 and all Australians were proud of your performance.[2]

Implicit in Cooley's letter was confirmation that Honeysuckle's days were numbered, its dish having been built specifically for the Apollo missions which were scheduled to conclude by 1973. Although still some way off, the end was also in sight for Orroral Valley with special space tracking satellites already on the drawing board. It was increasingly clear that the future of space tracking in Australia would centre on Tidbinbilla, that would come to be known as the Canberra Deep Space Communication Complex. In early 1969, the construction of a giant 210-foot dish had begun there. Made up of more than 1000 aluminium panels with a surface area of over 4000 square feet, this dish was three times the size of Honeysuckle's and six times more sensitive. It was designed to extend the useful life of a spacecraft travelling hundreds of millions of miles away from the Earth. Along with similar sized dishes at Goldstone and Madrid, it would complete NASA's deep space network. Headquartered at the Jet Propulsion Laboratory in Pasadena, California, this network would allow NASA to track objects in deep space around the clock. At the same time Tidbinbilla was fitted out with a brand-new telemetry system, to handle the surging number of long-range communications created by simultaneous, multiple missions. After the near faultless Apollo 12 mission in November 1969, it was agreed that NASA's most experienced station director in Australia, Tom Reid, would move over to Tidbinbilla to supervise its expansion, while its director, Don Gray, would move to Honeysuckle.[3]

This switch took place just before Apollo 13. For that mission, Tom found himself in charge of Honeysuckle's wing station. Just after 1 pm on 14 April 1970, about an hour after Apollo 13 came in view of the Australian tracking stations, an emergency was declared when an exploding oxygen tank crippled the spacecraft. The planned lunar landing was cancelled. Using a slingshot manoeuvre around the Moon, James Lovell, Jack Swigert and Fred Haise were able to limp home. Writing to Prime Minister Gorton, President Nixon said, 'Please convey my personal thanks to all of your people who worked so hard to maintain our communications with the weakened Apollo XIII spacecraft as it returned to Earth.' Even though it was Don Gray who led this magnificent effort it was Tom Reid's old Honeysuckle team that had followed through.[4]

To recognise Tom's work as Honeysuckle's station director, the Commonwealth government offered him membership of The Most Excellent Order of the British Empire which, in those days, was a gong received by Australians for 'prominent national achievement'. Tom was reluctant to accept because he felt that he was being honoured for just doing his job. But Margaret talked him around. On 24 April 1970, he presented himself to Queen Elizabeth at Government House, Canberra. While Her Majesty uttered a few special words to each recipient, it was different for Tom who ended up spending a couple of minutes answering the Queen's questions about Apollo 13.[5]

Tidbinbilla continued to act as Honeysuckle's wing station for the remainder of the Apollo missions, including Apollo 15 where a lunar rover was used for the first time. With a range of over 20 miles, it was capable of carrying two astronauts at a speed of 8 miles per hour. And it needed to be tracked as if it was a spacecraft in its own right. So, like the lunar module, it was rigged up with a small S-band antenna, to maintain links with Honeysuckle, Tidbinbilla, and the other tracking stations. Although Tidbinbilla's new 210-foot dish was not formally commissioned in time for Apollo 17, it was in working order. And Tom Reid managed to talk his NASA bosses into letting him rig it up to track this mission which ran for twelve days from 7 December 1972. On December 14, with Tidbinbilla in view, the last man on the Moon, Gene Cernan, climbed back onboard his lunar module. As

Cernan went to flick the yellow ignition switch to blast off back to the command module, he turned to his crewman, Jack Schmitt, and said, 'Okay Jack, now let's get off.' They were the last words spoken on the Moon.[6]

Following the end of the Apollo program, Tidbinbilla refocused on its primary purpose, communicating with man-made objects in deep space. These included Viking 1, the first attempt to land a spacecraft on Mars. After a ten-month journey to the Red Planet, Viking 1's lander had to be separated from its orbiter. This command and another, to commence the lander's descent, were given to its onboard computer by Tom Reid's Tidbinbilla team shortly after 8 pm on 21 July 1976 local time. Just over 3 hours later, the lander touched down safely, beginning what would be a surface mission lasting over 2300 days. Another, the Pioneer Multiprobe to Venus, involved a so-called 'bus' and the four probes it was carrying, separating and descending to that planet's surface. Again, Tom Reid's team played a key role in tracking, transmitting commands and receiving telemetry from each of these five space vehicles simultaneously, over a distance of 160 million miles. During the early stages of these missions, when Tidbinbilla's big dish couldn't be angled down far enough, Honeysuckle's little dish acted as an auxiliary antenna.[7]

The complex planning and coordination of these deep space missions required Tom to make regular trips to the Jet Propulsion Laboratory in Pasadena, California, so much so that he became a favourite among the flight attendants. Tongue-in-cheek, the executive vice president of the International Association of First Class Airline Stewardesses wrote to Tom's Australian boss.

> When word of any impending travel by Mr Reid is received by our organisation, special reports are immediately sent to all purveyors of alcoholic beverages throughout our international routes, alerting them to the fact that we will shortly be in need of an increased supply of their products on our Australia to US flights.[8]

One of the highlights of these trips for Tom was staying with his 'wee aunty' Jan, who having first migrated from Scotland to Canada with her husband Jack McLaren in 1952, had ended up not far from Pasadena in the City of Santa Clarita. Averaging three trips a year over twenty years, Tom would be picked up by Jan at the airport each time and driven at break-neck speed to the McLarens' home. Even though this little Scottish lady was barely able to see over the steering wheel, she would duck and weave through the heavy Los Angeles traffic like a native-born local. The McLarens' house had a view of the Sierra Pelona Mountains. Tom liked nothing more than chilling out for a few days to take in this view, as well as Jan's deep-fried fish and chips, a treat they had enjoyed as children while holidaying on the Scottish coast. Apart from buying presents for Margaret and the kids, Tom enjoyed reminiscing with Jan, although sometimes the discussions could be pointed. Having spent most of his adult life sporting a naval/ NASA short back and sides haircut, Tom turned up one day with his hair much longer and modishly styled. When Jan queried him, he replied, 'That's what you get when you marry a younger woman.'[9]

It was the Voyager program that best reflected the long-term nature of Tom Reid's work at Tidbinbilla. Indeed, Voyager continues right up to the present day, exploring interstellar space over 12 billion miles away. Launched in 1977 to take advantage of the favourable alignment of Jupiter, Saturn, Uranus and Neptune, Voyager had approached Jupiter in 1979. At that time, Tom's team needed to allow 40 minutes for commands received from Pasadena to reach Voyager. That's how long it took them to travel at the speed of light from Tidbinbilla's big dish to the spacecraft's onboard computers. The following year Voyager passed Saturn, and in 1986, Uranus. Messages now took more than 5 hours to travel almost 2 billion miles. And messages received at Tidbinbilla from Voyager as it passed Uranus were so weak that it would have taken twenty-four million years to gather enough strength from them to light a five-watt globe for a thousandth of a second. In

January 1986, Tom was interviewed by a *Canberra Times* journalist who reported:

> Tom Reid's window stares straight out on to the big white dish. The telephone on his desk is a direct line to space scientists in the United States and pinned to his walls are outer-space snapshots – the first pictures of Mars ('It's just like the desert at Woomera') … 'Voyager is a twelve-year mission,' he said, 'looking for the edge of the influence of the Sun.' Mr Reid, don't you sometimes feel stunned by all of this? 'Yes … sometimes we say "gee whiz" too.'[10]

During Tom Reid's tenure as Tidbinbilla's director, the other tracking stations were wound down and consolidated. Woomera's deep space facility was closed while Honeysuckle came under Tom's direct control in 1973. By 1981, deep cuts to NASA's budget necessitated Honeysuckle's complete closure. Its 85-foot dish was dismantled and transported to Tidbinbilla, where it was rebuilt and put to work as a supporting antenna. In 1985, Orroral Valley also closed. While its dish was given to the University of Tasmania for radio astronomy, some of its equipment ended up at Tidbinbilla. As these other stations closed, Tidbinbilla expanded. According to NASA historian Douglas Mudgway:

> Most of the major expansion…at Tidbinbilla took place during Tom Reid's eighteen-year directorship. The 210-foot antenna was built and later increased in size to 230 feet; the 85-foot antenna was enhanced to 110 feet; a new high-efficiency 110-foot antenna was built; and X-band uplinks and downlinks were added. Their successful integration into the network and subsequent record of outstanding service to NASA…owed much to his cooperation.[11]

Margaret Reid, meanwhile, had been elected president of the Liberal Party's Australian Capital Territory Division. Professing no ongoing parliamentary ambitions, Margaret was happy to support Senator John Knight, the Territory's only Liberal in the Federal Parliament. Knight had a bright future and so it came as a shock when he died suddenly in 1981, aged just thirty-seven. A number of candidates nominated for preselection to replace him, among them a former MP who had been defeated at the 1980 election. Because of Canberra's population demographic, Knight's old seat was as challenging as a marginal electorate. And many local Liberals believed that Margaret would be the best candidate but she was reluctant to put her hand up. With commitments to Tom and their family, Margaret was in two minds. However, knowing that she had sacrificed her parliamentary ambitions to marry him, Tom intervened. As Margaret recalled:

> My first reaction was not to stand. But none of the children were still at home and I thought about it. One morning, Tom said: 'You know you're the only one who can do his (Knight's) job'. And in a sense, deep down, that was what I wanted to hear.[12]

Having gone on to win the preselection convincingly, Margaret's election to the Senate was a constitutional formality. For the next couple of years as she did the Canberra rounds, Tom remained focused on Tidbinbilla, making exceptions for Anzac Day services and functions at the War Memorial. If people at such events happened to ask what he did, Tom simply said that he was 'a public servant', making no mention of NASA.[13]

In March 1983, the Liberal government was swept from office and it took more than two weeks of counting before Margaret knew that she had been returned. Following this close shave, she focused her ambitions on parliamentary rather than shadow ministerial roles, giving her plenty of time to work her Canberra constituency on non-sitting days. Even so, Margaret made time for her growing tribe of grandchildren, among other things taking them on tours of Tidbinbilla. The eldest, Nick's daughter Megan, later recalled:

> My earliest memories of Grandpa are, fittingly, from Tidbinbilla...I realised even at age four that Tidbinbilla was a big, important place, where Grandpa was an important man. His work was very serious. Very precise. I knew I needed to be quiet and well behaved. This, however, did not stop me from trying to press all the buttons in the control rooms while Grandpa was not looking. Needless to say, he was not amused and I was always whisked off to the canteen for some fish and chips and a can of Mellow Yellow.[14]

With the extensions to Tidbinbilla's big dish completed, Tom Reid stepped down as director on 14 October 1988. For Voyager's fly-past of Neptune, scheduled to take place in August the following year, Tom had bequeathed to his successor, Mike Dinn, a dish whose enlarged diameter had increased its reception capacity by fifty per cent. Weighing almost 4000 tons and large enough 'to throw a football field into', it could now pick up signals measured in billionths of a billionth of a watt. Two days before his retirement, Tom sent a telex to the worldwide space tracking network. It read in part:

> My association with NASA began in 1960 with Project Mercury...and during the intervening years, I have been fortunate to be involved in most of NASA's space programs... It has been stimulating to be involved, in a small way, in one of mankind's greatest endeavours...I would like to thank all of you...for making this possible and I acknowledge the great debt of gratitude I owe to all of the personnel at the Australian stations for the first-class support provided to me.

Rather than the usual sign off on this telex, Tom simply typed in 'Action? PURGE'. And on the line below that: 'Purged'. It was a typical light-hearted demonstration of the man's modesty. Equally light-hearted but nevertheless heartfelt was a poem dedicated to Tom by his executive officer, Bernie Scrivener. Entitled 'The Life and Times of a Glaswegian, a Chronology in Fractured Verse', it expressed the sentiments of the Australian space tracking community in twenty-five stanzas. Among them was this one.

> He'd been hard and tough but fair
> For a job there was to be done
> And the country would never have got there
> If he'd been soft and pallid and wan.

NASA's bigwigs also understood the immense debt they owed Tom Reid. In April 1989, the highest ranked American official ever to visit an Australian tracking station, the vice president of the United States, made a special trip all the way out to Tidbinbilla from the American Embassy in Canberra to invest Tom with NASA's Exceptional Public Service Medal. After explaining to the assembled gathering how Reid had made a lasting impact on NASA's success, the vice president invited Reid to reply. 'Thank you, Mr Vice President, for this great honour,' Reid said, then he stepped back from the microphone, illustrating his lifelong aversion to blowing his own trumpet. Although his career as a tracking station director was at an end, Tom remained a sought-after consultant to NASA's deep space tracking network for many years.[15]

Margaret Reid had been navigating the vagaries of the Senate with considerable success, rising through the ranks to become the Liberals' whip in 1987. Responsible for maintaining party discipline, Margaret was moderate in her views. In August 1996, she was elected president of the Senate. As such Margaret was the most senior parliamentary presiding officer in Australia and the first female to hold that office. Being head of the Department of the Senate meant that Margaret's executive responsibilities were similar to those of a government minister. For a time, she served as president of the Commonwealth Parliamentary Association and also supported parliamentary institutions in developing countries of the South Pacific. During her six years as president, Margaret travelled extensively on official business and almost invariably

Tom was by her side. He would read widely to prepare himself for these official trips, also taking the time to learn a little of the language of whichever country he and Margaret happened to be visiting. Having been a quintessential NASA alpha male for most of his working life, Tom was content to support Margaret in her equally impressive career. When being introduced to one former prime minister's wife who, like almost everyone else associated with politics, knew nothing of his time with NASA, he would introduce himself as Margaret's 'cleaner, ironer, cook and chief bottle washer'. On 6 February 2003, as Margaret prepared to depart the Senate, she gave her valedictory speech, recalling a conversation with one of the Labor Party's leading senators.

> The other day Robert Ray said to me, 'What date are you retiring?' I am not sure whether he was anxious that it should be sooner or later. I said, '14 February'. He said, 'Valentine's Day; it'll be the best Valentine's present Tom Reid has ever received.' I think it will be as well.[16]

Although Tom had been retired for some years his unique contribution to space tracking had not been forgotten in the United States. In January 2007, he and Margaret flew to Houston where, at a gala dinner, he was presented with a special award by the last man on the Moon, Gene Cernan. The award read: 'To Thomas Reid, for his leadership role in creating and operating "the global telemetry communications network that made communication with spacecraft possible".'[17]

Better than anyone, Tom understood the historic role that the Honeysuckle Creek and Orroral Valley tracking stations had played in the Space race and it upset him that their long-abandoned operations buildings had been gutted and defaced by vandals. Despite Margaret's attempts in the Senate to convince the ACT government to restore these buildings for use as offices by the Territory's Parks and Conservation Service, they were deemed to be beyond repair and eventually demolished. Today little is left of these tracking stations apart from their concrete foundations, crumbling bitumen laneways and the occasional piece of rusting structural steel. Thanks to foraging kangaroos there are also vestiges of the old campus-like lawns which have not yet been swallowed up by the surrounding bushland.

Honeysuckle's 85-foot dish remains intact. Having been moved to Tidbinbilla in 1982, it continued in service until 2010 when it was decommissioned because of metal fatigue in its structure and significant non-repairable wear in its drive mechanics. Recognising its significance as the little dish that had brought Neil Armstrong's first step to the world, the CSIRO which now operates Tidbinbilla on behalf of NASA, decided that it should be permanently preserved. Today it stands at the entrance to Tidbinbilla, with its dish facing upwards in a fixed and stowed position, a revered historic landmark. As for Tom Reid and Neil Armstrong, they met only once, at a Sydney luncheon for the Apollo 11 astronauts hosted by Prime Minister Gorton on 1 November 1969. What they discussed is unknown but they had more than their NASA experiences in common. With a partly Scottish ancestry, Armstrong's shyness, like Tom's, was often mistaken for aloofness. And like Tom, Armstrong was only ever fully understood by his immediate family. As one of Armstrong's closest NASA colleagues, Milt Thompson, once said of him, 'I knew him, but I didn't know him'; a comment many of the Honeysuckle team might have made about Tom too.[18]

Following Margaret Reid's retirement, she and Tom enjoyed seven wonderful years together, travelling and spending time with their grandchildren. To them Tom was known variously as Papa, Putti, Grandpa, Grandpaw and Granddog. In early 2010, he developed a form of lung cancer which the Veterans' Affairs Department accepted had its origins in the asbestos lagging used aboard HMAS *Warramunga*. Despite his devastating diagnosis, Tom carried on with great dignity and equanimity, astonishing all those who gathered around his hospital bed in his final hours by thanking his respiratory physician Dr Mark Hurwitz in a short but moving speech. Having arranged to donate his body to science, Tom Reid passed away on 2 October 2010, forty-five years to the day after the death of his first wife, Betty.[19]

In a heartfelt eulogy at Tom Reid's wake his elder daughter, Marg, had this to say:

> Dad was a man of amazing intellect, sharp mind and attention to detail...But he was understated and sadly I think that few of us really appreciate the amazing achievements he made... Dad left an enormous legacy and we are all incredibly proud of what a great man he was.[20]

Ironically Tom Reid would never have considered himself a great man. And if I had attempted to write this book during his lifetime, he would have done everything in his power to stop me. Tom's focus was always on the teams he led, not on himself. Nor was he carried away by the superpowers' space race. Yet his contribution to space tracking over almost three decades, which helped to make manned missions to the Moon and unmanned space flights beyond the solar system possible, was immense. Worldwide, there were hundreds of thousands of NASA employees and contractors, and hundreds of NASA leaders who also made wonderful contributions. There was, however, one achievement that made Tom Reid's contribution unique. That was his leadership in bringing a dysfunctional Apollo tracking station to a point of professional excellence where the dish at Honeysuckle Creek could do something it was not built to do, was not meant to do, and no other station was able to do: bring to the world live television of Neil Armstrong's first step and record for all time one of humanity's greatest achievements.

Acknowledgments

When I first approached my indefatigable literary agent, Lyn Tranter, with a proposal for a book about the man behind the dish at Honeysuckle Creek, her response was, 'Good idea. But how does a lawyer/politician like you get to know a rocket scientist?' After I told her I had dated Tom Reid's elder daughter, Marg, for three years and it had ended badly, she said, 'Right. I want three chapters of back story on that.' Realising an attempt at this would not be advisable without the consent of my wife, Kerry, and of Marg Reid, I floated Lyn's idea with each of them. Their responses were similar. As Kerry and I, and Marg and her husband Michael, have each been happily married for a very long time, there would be no problem. Although the original three chapter back story has been edited down and is now incorporated into the book's introduction, it still remains central to explaining Tom Reid as I remember him. My first big thanks are to Kerry and Marg.

Without the enthusiastic cooperation and support of the Honourable Margaret Reid AO, this book would not have been possible. In addition to spending many hours with me over a number of days talking about the Tom Reid she knew, I am very grateful to Margaret for giving me access to numerous documents, tapes and photographs chronicling Tom's life. Other members of the Reid family I would like to thank include Bob, Tom, Nick and Megan Reid, Danae Griffith and Jan McLaren; and from the McKenna family, Jim and John McKenna.

Most of Tom Reid's surviving colleagues from his Navy, Woomera and Honeysuckle Creek days are now in their late seventies and eighties. Because I am not tech-savvy, I must have tested their patience as they attempted to explain the workings of radars, dishes, transmitters, receivers and computers, all from the 1950s and 1960s. I am indebted to them for fielding my seemingly endless questions with good grace. They include: from the Royal Australian Navy, the late Rear Admiral Fred Lynam CBE; from the Goddard Space Flight Centre, George Harris Jr; from Woomera, Ken Anderson and Bill Miller; from Orroral Valley, Philip Clark; and from Honeysuckle Creek, Tony Cobden, John Crowe, Mike Dinn, Kevin Gallegos, Jim Kirkpatrick, Hamish Lindsay, John Saxon, Bernard Smith and Bryan Sullivan.

Although Colin Mackellar is not an engineer or technician and has never worked at a tracking station, Honeysuckle Creek's veterans acknowledge that he knows as much as they do, having spent literally years of his spare time developing a website dedicated to the station's history and workings: <www.honeysucklecreek.net>. Colin is held in such high regard by Honeysuckle's surviving engineers and technicians that they were all enthusiastic when I told them he would be my technical reader. I am deeply indebted to Colin for allowing me to draw upon stories from his website and for checking two drafts of my manuscript for technical errors. That said, any remaining mistakes are mine.

I would also like to acknowledge the assistance of Rear Admiral Simon Cullen AM RAN (ret'd), British naval historian Tim Dougall, Tony Eggleton AO CVO, Roger Kirchner and the Right Honourable Ian Sinclair AC; acknowledging also the advice of my friend and fellow writer, Charlotte Rogers Brown, whose editorial suggestions have been invaluable.

Finally, I want to thank Phillipa McGuinness and her wonderful team at NewSouth Publishing; my marvellous literary agent, Lyn Tranter and her husband, John; and Professor Peter

Radan and Professor Ben Schreer whose ongoing support for me in my role as an adjunct professor at Macquarie University is deeply appreciated.

Andrew Tink

Notes

INTRODUCTION

1 Hamish Lindsay, *Tracking Apollo to the Moon*, Springer, London, 2001, p. 226.
2 Lindsay, p. 227.
3 *Sydney Morning Herald*, 21 July 1969.
4 Lindsay, p. 235.
5 Lindsay, p. 239.
6 *Canberra Times*, 5 December 1973.
7 *Canberra Times*, 15 May and 29 November 1973.
8 The CSIRO's spokesperson at the Canberra Deep Space Communications Complex, Glen Nagle, recently acknowledged it was Honeysuckle Creek that had brought to the world the TV images of Neil Armstrong's first step: *The Guardian*, 14 August 2017; See also *The Monthly*, November 2017, Issue 139, pp. 16–18.
9 www.amazon.com/review/R202C656VM9L96/ref=cm_cr_rdp_perm (accessed 12 September 2017)
10 http://members.pcug.org.au/~mdinn/TheDish/ (accessed 1 October 2017).
11 http://members.pcug.org.au/~mdinn/TheDish/ (accessed 1 October 2017).

OCH, TOMMY!

1 Danae Reid's speech at Tom Reid's 70th birthday party.
2 Anna Blair, *Miss Cranston's Omnibus*, Lomond Books, Edinburgh, 1998, p. 342; Author's interview with Bob Reid 30 November 2016; Reference dated 2 October 1924 for Thomas Reid from D. Skiffington.
3 Berry, S and White, H (eds.), *Glasgow Observed*, John Donald Publishers, Edinburgh, 1987, p. 200; Bob Reid, 18 November 2016; Danae Reid's speech.
4 Nelson Sizer, *The Road to Wealth*, Lee and Shepard, Boston, 1882, p. 44.
5 Author's interview with the Honourable Margaret Reid AO (Margaret Reid) 1 April 2016; Berry and Whyte, pp. 209–10.
6 Author's interviews with Jan McLaren on 7 August 2016; Bob Reid, 18 November 2016.

7 Jan McLaren, 7 August 2016; www.theglasgowstory.com/image/?inum=TGSA01990 (accessed 1 December 2016); Blair, A., *Tea at Miss Cranston's*, Shepheard-Walwyn, London, 1988, pp. 165–66.
8 Thornwood Primary School class photo 1937.
9 Jan McLaren, 7 August 2016; Blair (1998); pp. 377–79.
10 Jan McLaren, 7 August 2016; Blair (1998); pp. 379.
11 Bob Reid, 7 December 2016; Jan McLaren, 7 August 2016; www.bbc.co.uk/scotland/aboutus/wirelesstoweb/history/ (accessed 7 December 2016).
12 Jules Verne, *From the Earth to the Moon and Around the Moon*, Wordsworth Editions, Hertfordshire, 2011, pp. vii, xv–xvi, 78, 192–195, 383–389; 406–414; www.oberth-museum.org/index_e.html (accessed 27 December 2016); www.nasa.gov/centers/goddard/about/history/dr_goddard.html (accessed 27 December 2016); *World News*, Sydney, 21 August 1920 and 20 November 1929; *Central Queensland Herald*, 21 November 1935; Jan McLaren, 7 August 2016; Danae Griffith's speech.
13 Jan McLaren, 7 August 2016; Bob Reid, 18 November 2016.

KEEP YOUR CHIN UP

1 Ben Wicks, *No Time to Wave Goodbye*, Bloomsbury Publishing, London, 1988, p. 23.
2 Berry and Whyte, p. 225; Bob Reid, 18 November 2016.
3 Bob Reid, 18 November 2016.
4 Bob Reid, 20 November 2016; Jan McLaren, 7 August 2016.
5 Bob Reid, 18 and 20 November 2016; Jan McLaren, 7 August 2016; John Williamson, *A History of Morrison's Academy, Crieff*, A.D. Garrie & Son, Auchterarder, 1980, p. 36.
6 Williamson, p. 38.
7 Bob Reid, 18 November 2016; www.armadale.org.uk/ardoch.htm (accessed 17 November 2016); www.castlefortsbattles.co.uk/perth_fife/ardoch (accessed 11 December 2016).
8 John Macleod, *River of Fire, The Clydebank Blitz*, Birlinn, Edinburgh, 2010, pp. 99, 100, 114, 161; Jan McLaren, 7 August 2016; Bob Reid, 20 November 2016; *Dudgeon*, pp. 255, 273.
9 Macleod, pp. 202–204; Bob Reid, 18 November 2016.
10 Williamson, pp. 38-39.
11 Barbara Oakley, *A Mind for Numbers*, Penguin, New York, 2014, pp. 11–13, 20; www.livescience.com/54370-math-brain-network-discovered.html (accessed 15 December 2016).
12 Oakley, p. 70.
13 Oakley, pp. 55, 119.
14 Oakley, p. 198; Perth and Kinross Education Committee, 1944, Leaving Certificate Examination, Mathematics Higher Grade (Second Paper) & Science Higher Grade (Physics).
15 Edward Frenkel, *Love and Math*, Basic Books, New York, 2013, pp. 2, 5, 23.
16 Bob Reid, 18 November 2016; Thomas Reid's Royal Navy service records; Paper dated 19 August 2016, explaining and amplifying Tom Reid's Royal Navy service records, prepared for the author by naval historian Tim Dougall, p. 3.

SENIOR SERVICE BRAND

1 Jan McLaren, 7 August 2016; Tom Reid's Royal Navy service records; Dougall, pp. 3–4.
2 Dougall, pp. 2–4; Tom Reid's Royal Navy service records.
3 Dougall, p. 5; Email dated 17 January 2017 from Tim Dougall to the author.
4 Dougall, pp. 8–9.
5 *Popular Science*, December 1941, p. 56; Derek Howse, *Radar at Sea, the Royal Navy in World War 2*, Naval Institute Press, Annapolis, Maryland, 1993, p. 201.
6 Dougall, pp. 9–10.
7 Howse, pp. 247–48.
8 Dougall, p. 10.
9 Bob Reid, 18 November 2016; Dougall, p. 12.
10 Dougall, pp. 12–13; www.royalnavyresearcharchive.org.uk/BFP-EIF/Ships/GOLDEN_HIND.htm (accessed 14 January 2017); Cumberland Argus, 21 November 1945; Author's interview with Honeysuckle Creek historian Colin Mackellar, 11 March 2016; West Australian, 9 July 1946.
11 Dougall, pp. 16–20.
12 Dougall, pp. 2, 18–20; Tom Reid's Navy service records.
13 Author's interview with Bob Reid, 18 November 2016; www.scottisharchitects.org.uk/building_full.php?id=200466 (accessed 16 January 2017).
14 Hector Hetherington (ed.) *Fortuna Domus*, University of Glasgow, Glasgow, 1952, pp. 108–112.
15 *Fortuna Domus*, pp. 335, 353–54.
16 University of Glasgow, School of Electrical Engineering, Class of 1952 photo; Faculty of Engineering, *Faculty Handbook*, University of Glasgow, n/d, p. 26.
17 P. A. Bate, 'Bernard Hague (1893–1960)', *The Galpin Society Journal*, Volume 15 (March 1962) pp. 92–94; B. Hague, *An Introduction to Vector Analysis: for Physicists and Engineers*, Methuen, London, 1939, p. ix.
18 Various letters and certificates to Tom Reid from the University of Glasgow, issued between 1948 and 1952; Bob Reid, 19 November 2016.

LIEUTENANT SPARKS

1 Bob Reid, 19 November 2016; Jan McLaren, 7 August 2016.
2 www.johnwollf.id.au/calculators/comptometers/Operator.htm accessed 28 January 2017.
3 Bob Reid, 19 November 2016; Dudgeon, p. 30.
4 Tom Reid's Royal Navy service records; Bob Reid, 19 November 2016.
5 Bob Reid, 19 November 2016.
6 Bob Reid, 18 and 19 November 2016.
7 Bob Reid, 19 November 2016.
8 Email dated 23 November 2016 from John McKenna to the author; www.medicalnewstoday.com/articles/8856.php (accessed 30 January 2017).
9 Letter dated 29 April 1952 from Professor Bernard Hague to Thomas Reid.
10 *Faculty of Engineering Final Year Book 1948–1952*.
11 Bob Reid, 19 November 2016; Author's interview with Marg Reid, 27 September 2016; Jan McLaren, 7 August 2016.
12 Marg Reid, 27 September 2016; Marg Reid's speech at her father's wake, October 2010.

13 Thomas and Elizabeth Reid's Marriage Certificate; Tom Reid's RAN service record, NAA: A6769, Reid T (accessed 4 March 2016).
14 *Canberra Times*, 5 November 1952; http://korean-war.commemoration.gov.au (accessed 1 February 2017); Peter Firkins, *Of Nautilus and Eagles*, Cassell Australia, Sydney, 1975, pp. 227–28; www.navy.gov.au/hmas-condamine (accessed 3 February 2017).
15 *The Adelaide Mail*, 23 March 1940; www.navy.gov.au/biography/rear-admiral-daryall-frederick-lynam (accessed 3 February 2016).
16 Author's interview with Rear Admiral Fred Lynam CB, RAN (Ret'd), 14 October 2016.
17 *The Melbourne Age*, 9 August 1941; Marg Reid, 27 September 2016; Bob Reid, 19 November 2016.
18 www.navy.gov.au/hmas-vengeance (accessed 3 February 2017); Fred Lynam, 14 October 2016; http://adb.anu.edu.au/biography/becher-otto-humphrey-9465 (accessed 3 February 2017); Tom Reid's Certificate of Service, HMAS *Vengeance*.
19 Reid family photos 1954–1956; Tom Reid's Certificates of Service, HMAS *Leeuwin*.
20 www.navy.gov.au/hmas-warramunga-i (accessed 4 February 2017).
21 Fred Lynam, 14 October 2016; Author's interview with Rear Admiral Simon Cullen, AM, RAN (Ret'd), 15 February 2017.
22 Fred Lynam, 14 October 2016; Bob Reid, 19 November 2016.
23 *The Argus*, Melbourne, 27 November and 4 December 1956; *The Canberra Times*, 8 March 1957; http://corporate.olympics.com.au/games/melbourne (accessed 4 February 2017); www.navy.gov.au/hmas-warramunga-i (accessed 4 February 2017); Tom Reid's Certificate of Service, HMAS *Warramunga*.

WOOMERA

1 'Woomera...a home on the range', *Australian Women's Weekly*, 3 December 1958.
2 Ivan Southall, *Woomera*, Angus and Robertson, Sydney, 1962, pp. 1–2; Peter Morton, *Fire Across the Desert*, Australian Government Publishing Service, Canberra, 1989, p. 125.
3 Southall, pp. 74, 80; Morton, pp. 228, 231–32; Marg Reid, 27 September 2016; 'Woomera...a home on the range', *Australian Women's Weekly*.
4 Southall, pp. 74–75; Morton, pp. 238–43.
5 Morton, pp. 3-9.
6 Southall, pp. 12–13, 46–47; Morton, pp. 15–16, 213.
7 Morton, pp. 102, 223, 261; Southall, pp. 49, 90–91.
8 Southall, pp. 136–37; Colin Mackellar, 19 September 2017.
9 Southall, pp. 138–39.
10 Biographical notes and related documents concerning Tom Reid; Southall, pp. 137, 139.
11 Southall, pp. 215, 292; Biographical notes; Fred Lynam, 14 October 2016.
12 Author's interview with Bill Miller 9 June 2016; John Blaxland and Rhys Crawley, *The Secret Cold War, The Official History of ASIO, 1975–1989*, Volume III, Allen and Unwin, Sydney, 2016, p. 241.
13 Southall, pp. 54–59; Morton, pp. 409–18.

14 Morton, pp. 418–20.
15 Southall, pp. 241–43, Morton, p. 423.
16 Morton, pp. 288, 294–97; Southall, p. 135.
17 Southall, p. 109.
18 Marg Reid, 27 September 2016; Southall, p. 79; Bill Miller, 9 June 2016; Morton, p. 101.

SPUTNIK AND *KAPUTNIK*

1 Lloyd S. Swenson, James M. Grimwood, Charles C. Alexander, *This New Ocean: a history of Project Mercury*, Red and Black Publishers, St Petersburg, Florida, originally published in 1966, p. 25.
2 Swenson, pp. 17–25; Sunny Tsiao, *Read You Loud and Clear; the story of NASA's Spaceflight and Data Tracking Network*, The NASA History Series, Washington DC, 2008, p. 9.
3 Tsiao, pp. xxxiii, 13.
4 Tsiao, pp. 13–14, 19, 21, 23.
5 Tsiao, p. 23.
6 Tsiao, p. 23; Lindsay, p. 15; Eugene Kranz, *Failure is not an Option*, Berkley Books, New York, 2001, p. 15; Roger D. Launius, *Sputnik and the Origins of the Space Age*, www.history.nasa.gov/sputnik/sputorig.html (accessed 1 November 2016).
7 Lindsay, p. 17; Tsiao, p. 24.
8 Walter A. McDougall, *The Heavens and the Earth a Political History of the Space Age*, Basic Books, New York, 1985, p. 122.
9 T. Keith Glennan, www.history.nasa.gov/Biographies/glennan.html (accessed 1 November 2016); Tsiao, p. ix; Lindsay, pp. 19–20.
10 Tsiao, pp. 70–73; John Catchpole, *Project Mercury*, Springer, Chichester UK, 2001, p. 119.
11 Tsiao, pp. 72, 92, 95.
12 Tsiao, pp. 74, 95; Author's interview with Ken, 23 November 2016.
13 Tsiao, p. 45; Catchpole, p. 124.
14 Catchpole, pp. 124–25; Morton, p. 256; Tsiao, p. 100; www.theqantassource.com/history.html (accessed 3 November 2106); George Harris Jr's email dated 23 March 2016 to the author.
15 Morton, p. 235; Marg Reid, 27 September 2016; Bill Miller, 9 June 2016.
16 *Introduction to Project Mercury and Site Handbook*, Western Electric Company, New York, NY, 1960, pp. 6–41, 6–42; www.honeysucklecreek.net/other_stations/red_lake (accessed 13 April 2016); Ken Anderson, 23 November 2016.
17 Bill Miller, 9 June and 26 October 2016.
18 *Mercury Handbook MG-101-2 1-4*; Tsiao, pp. 70, 97.
19 Lindsay, pp. 46, 52–53; Swenson, p. 318.

BEFORE THIS DECADE IS OUT

1 Lindsay, p. 21.
2 www.jfklibrary.org/JFK/JFK-in-History/The-Bay-of-Pigs.aspx (accessed 4 November 2016); Lindsay, p. 21.
3 Lindsay, pp. 21–22.
4 Lindsay, pp. 54–63.

5 Catchpole, p. 125.
6 Southall, pp. 225–26.
7 *Mercury Handbook*, 6–10; Kranz, pp. 61–62; Email dated 6 May 2016 from Gene Kranz to Colin Mackellar.
8 *Canberra Times*, 12, 13 April 1961; Catchpole, pp. 95, 99–100; Kranz, pp. 52–53.
9 Lindsay, pp. 64–68; Swenson, pp. 374–81; 'The Horrible Thing That Happened to Enos the Chimp When He Orbited Earth 50 Years Ago', *The Atlantic*, 29 November, 2011; Bill Miller, 9 June 2016; Email dated 6 May 2016 from Gene Kranz to Colin Mackellar; Email dated 16 February 2017 from Edwin Overholt Jr to Colin Mackellar; Tsiao, p. 101; Catchpole, pp. 315–16.
10 Email dated 1 August 2010 from Tom Reid to Colin Mackellar.
11 Danae Griffith's speech; Bill Miller, 9 June and 13 October 2016.
12 Lindsay, p. 68; Swenson, pp. 383–92; *Gibber Gabber*, Volume IX, Number 47; Catchpole, p. 321.
13 Marg Reid, 27 September 2016, together with Reid family photos.
14 *Mercury Handbook*, pp. 6–7; Bill Miller, 9 June and 26 October 2016; Marg Reid, 26 October 2016; Email dated 6 May 2016 from Gene Kranz to Colin Mackellar.
15 Tom Reid's interview with NASA historian Craig Waff on 5 September 1990.
16 Catchpole, pp. 125, 325.
17 Recording of *Friendship 7*'s full pass over Muchea and Red Lake on Orbit 1 located at www.honeysucklecreek.net/other_stations/muchea/friendship_7 (accessed on 10 November 2016); Bill Miller, 9 June and 26 October 2016; Ken Anderson, 23 November 2016.
18 Kranz, pp. 67–69; Catchpole, pp. 326–37; Swenson, p. 398.
19 Catchpole, pp. 329–30; Swenson, p. 398.
20 Swenson, pp. 398, 400, 402; Tsiao, p. xxxi; Catchpole, pp. 329–33.
21 Bill Miller, 26 October 2016; Catchpole, pp. 333–34.

ACADEMIC INTERLUDE

1 S.I. Evans, *My Years as Director 1961–1978*, The South Australian Institute of Technology, Adelaide, 1978, Foreword, pp. 7, 21; Annely Aeuckens, *The People's University*, South Australian Institute of Technology, Adelaide, 1989, pp. 189–90, 211, 214.
2 Evans, pp. 8, 9, 11, 12, 44; Aeuckens, p. 240.
3 Tom Reid's Curriculum Vitae.
4 Evans, pp. 26, 112.
5 Reid family photos.
6 Lindsay, pp. 87–89; Tsiao, pp. 105–07.
7 Philip Clark, *Acquisition*, Elect Printing, Canberra, 2012, pp. 13–14.
8 Clark, p. 14.
9 Clark, pp.14, 61.
10 www.honeysucklecreek.net/other_stations/tidbinbilla/tidbinbilla_early_days.html (accessed 20 February2017); Tsiao, p. 45.
11 Clark, pp. 1, 3, 17, 95; www.eoas.info/blogs/P003687b.htm (accessed 6 October 2016).
12 Clark, pp. 17, 240; Evans, p. 44; www.honeysucklecreek.net/other_stations/

orroral/index.html (accessed 22 February 2017).
13 Author's interview with Nick Reid, 31 May 2016; Author's interview with John McKenna, 26 November 2016.
14 National Capital Development Commission, *A Report on the Development of Canberra for the Five Year Period July 1962 – June 1967*, Canberra, 1962, pp. 9, 13.
15 *Canberra Times*, 4 September 1964.
16 *Canberra Times*, 10 March 1965.
17 Clark, p. 17.
18 *Canberra Times*, 28 October, 3 and 4 November 1964.

CASTLES IN THE AIR

1 Clark, p. 18.
2 Author's interview with Roger Kirchner, 3 March 2016; Clark, pp. 17, 19.
3 John McKenna, 26 November 2016; Letter dated 4 August 1965 from Betty Reid to her mother.
4 Alan Fitzgerald, *Fitzgerald's Canberra*, Dalton Publishing, Canberra, 1969, p. 7; Nick Reid, 31 May 2016; Betty Reid letter.
5 *Clark*, pp. 21, 59; Betty Reid letter.
6 Betty Reid letter; Reid family photos; Nick Reid, 31 May 2016; Author's interview with Tom Reid 11 April 2017.
7 Clark, pp. 27–28, 67; *Canberra Times*, 4 September 1965.
8 *Canberra Times*, 1 October 1965.
9 Clark, p. 67; *Canberra Times*, 2 October 1965.
10 Marg Reid, 30 March 2017; *Canberra Times*, 1 October 1965.
11 Canberra Hospital's clinical and nursing notes, dated 1 and 2 October 1965, relating to Elizabeth Reid.
12 Clark, p. 67; Nick Reid, 31 May 2016; Marg Reid, 30 March 2017; Tom Reid, 11 April 2017.
13 Clinical notes; Elizabeth Reid's Death Certificate.
14 *Canberra Times*, 4 October 1965.
15 Reid family photos; Bill Miller, 9 June 2016; Fred Lynam, 14 October 2016.
16 Marg Reid, 27 September 2016; Bob Reid, 18 November 2016.
17 Margaret Reid, 29 April 2016; *Canberra Times*, 14 February 1985.
18 Margaret Reid, 29 April 2016; Nick Reid, 31 May 2016; Clark, pp. 35–36.
19 *Canberra Times*, 4 September and 2 October 1965, 25 February 1966; Tsiao, p. 47.
20 Clark, pp. 112–13.
21 *Canberra Times*, 19 November 1965.
22 *Canberra Times*, 19 November 1965.

AT FIRST SIGHT

1 Clark, pp. 21, 79, 81.
2 Clark, pp. 71, 72; *Canberra Times*, 14 April 1966.
3 Clark, p. 71.
4 *Canberra Times*, 25 February 1966.
5 http://bioguide.congress.gov/scripts/biodisplay.pl?index=M000727 (accessed 17 March 2017); Colin Mackellar, 19 September 2017.
6 *Canberra Times*, 24 February 1966.

7 *Canberra Times*, 23 and 25 February 1966; Clark, p. 239.
8 Clark, p. 74.
9 Clark, pp. 22, 390; Author's interview with Philip Clark, 6 May 2017.
10 Clark, pp. 193, 197, 350; Margaret Reid, 29 April 2016.
11 Margaret Reid, 1 April 2016.
12 Margaret Reid, 4 April 2017.
13 Margaret Reid, 4 April 2017.
14 Margaret Reid, 1 April and 27 May 2016, 4 April 2017.
15 Margaret Reid, 29 April 2016; Author's interview with Chris Crowe, 10 April 2017; www.honeysucklecrek.net/people/songs.html (accessed 15 January 2018).
16 Margaret Reid, 4 April 2017.
17 Certificate of Marriage between Tom Reid and Margaret McLachlan on 25 February 1967.

A MOON MADE OF CHEESE?

1 *Canberra Times*, 14 April and 1 December 1966.
2 Tsiao, pp. 143–144.
3 Tsiao, p. 145; Lindsay, p. 149; Richard Starchurski, *Below Tranquility Base*, CreateSpace Independent Publishing Platform, North Charleston, 2013, pp. 312–14.
4 Tsiao, pp. 147–49.
5 Starchurski, pp. 301–12.
6 *Canberra Times*, 1 December 1966 and 28 January 1967.
7 *Canberra Times*, 1 December 1966; Kranz; p. 193.
8 *Canberra Times*, 4 and 7 January 1967; Kranz, p. 15; Lindsay, pp. 21, 53; www.honeysucklecreek.net/people/stories.html (accessed 13 September 2017).
9 *Canberra Times*, 12 January 1967.
10 Kranz, pp. 196–97, 204–05, 207; Lindsay, p. 155.
11 *Canberra Times*, 18 March 1967; *Good Neighbour (ACT)*, 1 April 1967; Lindsay, p. 325.
12 Kranz, p. 234.
13 Kranz, p. 206.
14 Colin Mackellar, 11 March 2016; Author's interview with John Crowe, 21 April 2017.
15 Colin Mackellar, 11 March 2016; www.honeysucklecreek.net/people/songs.html (accessed 15 January 2018).
16 John Crowe, 21 April 2017; Tsiao, p. 76; *Canberra Times*, 26 June, 3 July and 30 September 1967.
17 *Canberra Times*, 8 July 1967; Colin Mackellar, 11 March 2016; John Crowe, 21 April 2017; Starchurski, p. 304.
18 Colin Mackellar, 11 March 2016; John Crowe, 21 April 2017; Author's interview with Hamish Lindsay, 1 June 2016; Author's interview with John Saxon, 13 June 2016; Author's interview with Mike Dinn, 2 May 2016.
19 W.R.E. Notice 67/63, 4 August 1967.

A HARD MAN

1 Tom Reid's interview.
2 Starchurski, pp. 302–03.
3 Tom Reid's interview; Hamish Lindsay, 1 June 2016.
4 Mike Dinn, 2 May 2016; Starchurski, p. 312.
5 Author's interview with Jim Kirkpartick, 31 May 2016; Mike Dinn, 2 May 2016.
6 Jim Kirkpatrick, 31 May 2016; Mike Dinn, 2 May 2016.
7 Author's interview with Kevin Gallegos, 8 June 2016; Mike Dinn, 2 May 2016; Starchurski, p. 311; Lindsay, p. 189.
8 Starchurski, pp. 311–12; Mike Dinn, 2 May 2016.
9 Author's interview with Bryan Sullivan, 14 June 2016; Memo dated 6 September 1967 from John South to Ozro Covington.
10 Hamish Lindsay, 1 June 2016; Jim Kirkpatrick, 31 May 2016.
11 John Saxon, 13 June 2016; Author's interview with Tony Cobden, 12 May 2016; Kevin Gallegos, 8 June 2016; Hamish Lindsay, 1 June 2016; www.honeysucklecreek.net/early_days/index (accessed 5 January 2018).
12 Kevin Gallegos, 8 June 2016; John Saxon, 13 June 2016.
13 Margaret Reid, 1 April 2016; Danae Griffith's speech; *Canberra Times*, 6 August 1978.
14 Margaret Reid, 1 May and 6 June 2017.
15 Email dated 9 June 2017 from Marg Reid to the author.
16 Danae Griffith's speech.

SANTA CLAUS

1 John Crowe, 11 April 2017.
2 Bryan Sullivan, 14 June 2016.
3 'Buzz Phrase, Buzz Word Generator'.
4 Senator Gary Humphries, 'Mr Thomas Reid MBE', Adjournment Debate, Australian Senate, 16 November 2010; Mike Dinn, 2 May 2017; Jim Kirkpatrick, 31 May 2016; Bryan Sullivan, 14 June 2016; Lindsay, p. 165.
5 Bryan Sullivan, 14 June 2016.
6 John Saxon, 13 June 2016; Bryan Sullivan, 14 June 2016; Lindsay, p. 165.
7 Lindsay, pp. 167–71; John Saxon, 26 April 2017; *Canberra Times*, 8 November 1967.
8 Jim Kirkpatrick, 3 May 2016; Lindsay, pp. 166–67, 170.
9 John Saxon, 13 June 2016; Mike Dinn, 2 May 2016; Bryan Sullivan and Jackie French, *To The Moon and Back*, Angus and Robertson, Sydney, 2004, p. 45.
10 Lindsay, pp. 171–74; Kranz, pp. 225–26.
11 Lindsay, pp. 174–77; Dwight Steven-Boniecki, *Live TV From the Moon*, Apogee Books, Burlington Ontario, 2010, pp. 27–29.
12 Steven-Boniecki, pp. 57–59.
13 Steven-Boniecki, pp. 62–64.
14 *Canberra Times*, 22 October 1968; Lindsay, pp. 178–79.
15 Lindsay, pp. 178, 180: www.honeysucklecreek.net/people/stories.html (accessed 13 September 2017).
16 Letter dated 3 December 1968 from Bob Leslie to Tom Reid.
17 Mike Dinn's eulogy at Don Gray's funeral, 14 November 2014,

www.honeysucklecreek.net/people/don_gray (accessed 6 May 2017); *Canberra Times*, 16 and 23 December 2017.
18 Bryan Sullivan, 14 June 2016; Lindsay, p. 184.
19 Lindsay, pp. 179, 188, 193; *Canberra Times*, 26 December 1968.
20 Steven-Boniecki, pp. 68, 71, 73.
21 *Canberra Times*, 26 December 1968.

THE MOP HANDLE

1 Lindsay, pp. 193, 195, 198; Kranz, pp. 237, 246–247.
2 *Canberra Times*, 7 January 1969.
3 Sullivan and French, p. 67; *Canberra Times*, 31 December 1968; www.honeysucklecreek.net/msfn_missions/Apollo_8_party (accessed 12 September 2017).
4 *Canberra Times*, 7 January 1969.
5 Lindsay, pp. 196–97.
6 Kranz, p. 252; *Canberra Times*, 4 March 1969; Lindsay, pp. 194–97; Sullivan and French, p. 73; Robertson, *John Bolton and a New Window on the Universe*, NewSouth Publishing, Sydney, 2017, p. 309.
7 Email dated 20 September 2017 to the author from Colin Mackellar enclosing excerpts from Howard Klye's trip report, April 16–May 4 1969.
8 Sullivan and French, p. 75.
9 Lindsay, pp. 198–201; *Canberra Times*, 15 May 1969; Kranz, p. 255.
10 Lindsay, p. 205.
11 Mike Dinn, 12 May 2017.
12 Mike Dinn, 12 May 2017; D. Goddard and D. Milne (eds), *Parkes Thirty Years of Radio Astronomy*, CSIRO Publishing, Melbourne, 1994, p. 134; Tsiao, p. 174; Robinson, pp. 306–07.
13 Mike Dinn, 12 May 2017.
14 Lindsay, p. 211.

THE UPSIDE-DOWN CAMERA

1 Steven-Boniecki, pp. 105, 108.
2 Steven-Boniecki, pp. 106, 111.
3 Steven-Boniecki, pp. 19, 137.
4 Steven-Boniecki, pp. 27–29.
5 www.honeysucklecreek.net/Apollo_11/TV_from_Moon (accessed 3 July 2016).
6 Mike Dinn, 2 May 2016.
7 Mike Dinn, 2 May 2016.
8 Lindsay, p. 204.
9 Steven-Boniecki, p. 137; Bill Wood, *Apollo Television*, a paper privately published in 2005, p. 19.
10 Lindsay, pp. 193–94.
11 Kranz, p. 215; Lindsay, p. 152; Steven-Boniecki, pp. 115, 123–25, 149–50.
12 Lindsay, p. 154; Starchurski, p. 328; Kranz, p. 215; Email dated 13 March 2006 from Stan Lebar to Colin Mackellar.
13 Steven-Boniecki, pp. 121, 138, 145, 147.
14 *Honeysuckle Creek Permission TWX*, www.honeysucklecreek.net/msfn_missions/Apollo_11_mission (accessed 14 April 2016).

15 Sullivan and French, p. 85; Good Neighbour (ACT) 1 July 1969.
16 Lindsay, pp. 167–68; Colin Mackellar, 19 September 2017.
17 Lindsay, p. 214; *Canberra Times*, 17 July 1969.
18 Lindsay, p. 218; Starchurski, p. 218; Tsiao, p. 175.

LIKE FLIES TO A PICNIC LUNCH

1 Danae Griffith's speech.
2 Starchuski, p. 29.
3 Lindsay, p. 198.
4 Photo of Tom Reid and his father taken around 1939.
5 John Saxon, 13 June 2016.
6 Mike Dinn, 2 May 2016.
7 Sullivan and French, pp. 50–51, 75; *Canberra Times*, 21 July 1969; Jim Kirkpatrick, 31 May 2016; www.honeysucklecreek.net/people/stories.html (acessed 13 September 2017).
8 Mike Dinn, 2 May 2016.
9 Various photos of Tom Reid's office on the Honeysuckle Tracking Station website: www.honeysucklecreek.net (accessed on 20 June 2016); Starchurski, pp. 311–12.
10 Starchurski, pp. 18, 235–36, 313–15.
11 Starchurski, pp. 44, 312–13.
12 Kranz, pp. 13, 180, and photograph section; Starchurski, pp. 233, 235, 276; Chaikin, p. 190.
13 Starchurski, p. 233.
14 Starchurski, pp. 239–262; Kranz, pp. 284–85; Mike Dinn, 2 May 2016.

THE *EAGLE* HAS LANDED

1 Starchurski, pp. 4–6, 248; Kranz, p. 285.
2 Starchurski, pp. 248–49, 254, 312; Kranz, p. 287.
3 Starchurski, pp. 254–55; Email dated 8 February 2018 from Colin Mackellar to the author; www.honeysucklecreek.net/images/msfn_images/MCC-H/MCC_3 (accessed 9 February 2018); www.honeysucklecreek.net/audio/interviews/Jack_Garman_2_A11.mp3 (accessed 9 February 2018).
4 Starchurski, p. 261; Kranz, p. 291; Andrew Chaikin, *A Man on the Moon*, Penguin, New York, 2007, pp. 198–99; Neil Armstrong, *A Life of Flight*, St. Martin's Press, 2014, pp. 254–58; Colin Mackellar on 20 September 2017.
5 Chaikin, p. 197.
6 Starchurski, pp. 240–41, 246–47; Kranz, pp. 292–93.
7 Starchurski, p. 263; Chaikin, p. 201.
8 Lindsay, p. 230; Bryan Sullivan, 14 June 2016.
9 Margaret Reid, 1 April 2016.
10 Bryan Sullivan, 14 June 2016.
11 Kranz, p. 14.
12 Hamish Lindsay, 1 June 2016; Bryan Sullivan, 14 June 2016.
13 Bryan Sullivan, 14 June 2016.
14 Bryan Sullivan, 14 June 2016.
15 Kranz, pp. 257, 294; Starchurski, pp. 44–45.

16 www.jsc.nasa.gov/Bios/htmlbios/mccandless (accessed 5 July 2016); Starchurski, p. 127.
17 Kranz, pp. 189, 293; *Sydney Morning Herald*, 21 July 1969; Starchurski, pp. 263–64; Colin Mackellar, 20 September 2017.
18 Starchurski, p. 263.

A PRIME MINISTER 'BLINDED BY SCIENCE'

1 Chaikin, p. 204.
2 www.honeysucklecreek.net/station/ops_areas (accessed 5 July 2016); John Saxon, 13 June 2016; Bryan Sullivan; 14 June 2016.
3 Bryan Sullivan, 14 June 2016.
4 www.honeysucklecreek.net/Apollo_11/Gorton__ABC-TV (accessed 6 July 2016); Email dated 6 July 2016 from Tony Eggleton AO CVO to the author.
5 Margaret Reid, 27 May 2016; Tony Eggleton's email.
6 www.honeysucklecreek.net/Apollo_11/Gorton_ABC-TV (accessed 6 July 2016); Margaret Reid, 27 May 2016.
7 Bryan Sullivan 14 June 2016; www.honeysucklecreek.net/Apollo_11/Gorton_ABC-TV (accessed 6 July 2016); Margaret Reid, 27 May 2016.
8 John Saxon, 13 June 2016.
9 www.hq.nasa.gov/alsj/EdVonR_Apollo_Super8, www.honeysucklecreek.net/Apollo_11/Gorton_ABC-TV, www.honeysucklecreek.net/audio/interviews/Ed_Von_Renouard_part_1, (all accessed on 7 July 2016).
10 www.honeysucklecreek.net/other_stations/goldstone/index (accessed 7 July 2016).
11 Ian Hancock, *He Did It His Way*, Hodder, Sydney, 2002, pp. 174–79, 224–25; www.honeysucklecreek.net/Apollo_11/Gorton_Tour_of_HSK (accessed 7 July 2016); Tony Eggleton's email.
12 Bryan Sullivan, 14 June 2016.
13 www.honeysucklecreek.net/people/dinn (accessed 7 July 2016); Mike Dinn, 2 May 2016.
14 Sullivan and French, pp. 111–13; Lindsay, p. 230.

GOD DAMN IT: WE WERE READY!

1 www.tccs.act.gov.au/parks-consevation/parks-and-reserves (accessed 8 July 2016).
2 Hamish Lindsay, 1 June 2016; Lindsay, pp. 230–31.
3 Starchurski, pp. 30, 212.
4 Sullivan and French, p. 59.
5 www.hightechscience.org/apollo_spacesuit (accessed 9 July 2016); Chaikin, pp. 206–07.
6 Lindsay, p. 231.
7 Mike Dinn, 2 May 2016 and 5 October 2017.
8 www.honeysucklecreek.net/station/ops_areas (accessed 10 July 2016); Mike Dinn, 2 May 2016.
9 www.honeysucklecreek.net/people/saxon (accessed 10 July 2016).
10 John Saxon, 13 June 2016.
11 Author's email exchange on 1 May 2016 with The Right Honourable Ian Sinclair AC; Margaret Reid, 27 May 2016; Tom Reid, 11 April 2017.
12 John Saxon, 13 June 2016; Tsiao, p. 176; Colin Mackellar, 19 September 2017

and 18 May 2018.
13 www.honeysucklecreek.net/Apollo_11/PKS_and_HSK (accessed 11 July 2016).
14 Lindsay, pp. 233–34; Mike Dinn, 16 October 2017.
15 Email dated 2 May 2007 from Stan Lebar to Colin Mackellar.
16 John Saxon, 13 June 2016; Mike Dinn, 2 May 2016; Lindsay, p. 233; Email dated 13 March 2006 from Stan Lebar to Colin Mackellar; Colin Mackellar, 19 September 2017.
17 Lindsay, p. 232; www.honeysucklecreek.net/Apollo_11/comms_transcript (accessed 11 July 2016); Hamish Lindsay, 19 September 2017; Email dated 28 April 2009 from Ed Tarkington to Colin Mackellar.
18 Lindsay, p. 234; Colin Mackellar, 19 September 2017.
19 Bryan Sullivan, 14 June 2016.
20 Lindsay, pp. 234–35; Tsiao, p. 178.
21 Lindsay, p. 233.
22 www.honeysucklecreek.net/Apollo_11/comms_transcript (accessed 11 July 2016); Email dated 2 May 2007 from Stan Lebar to Colin Mackellar.
23 www.honeysucklecreek.net/Apollo_11/comms_transcript (accessed 11 July 2016); Margaret Reid, 27 May 2016; Bob Reid, 20 November 2016. www.honeysucklecreek.net/video/Tom_Reid_A11_20th.sm.mp4 (accessed 11 July 2016).

EPILOGUE

1 Kranz, p. 294; www.honeysucklecreek.net/Apollo_11/PKS_and_HSK (accessed 15 March 2016).
2 Letter dated 25 July 1969 from Alan Cooley to Tom Reid.
3 Clark, p. 221; *Canberra Times*, 14 December 1968; www.cdscc.nasa.gov/Pages/cdscc_history (accessed 3 June 2017); Lindsay, pp. 256–64; Michael Johnson, *Mission Control*, University Press of Florida, Gainesville, 2015, p. 146.
4 *Canberra Times*, 15 April 1970; Lindsay, pp. 265–93; Colin Mackellar, 19 September 2017.
5 Letter dated 6 April 1970 from the Governor-General's Official Secretary to Thomas Reid MBE; Margaret Reid, 29 April 1970.
6 Lindsay, pp. 309–10, 328, 333, 338; Email dated 4 June 2017 to the author from Colin Mackellar. At one time, Gene Cernan believed he had said, 'Ok Jack, let's get this mother out of here'. But in the relevant audio tape at 188:01:25, he is clearly heard saying, 'Ok Jack, let's get off': Colin Mackellar, 19 September 2017.
7 *Canberra Times*, 3, 16, 20 July 1976 and 22 May 1978; Johnson, p. 146.
8 Letter dated 19 May 1977 from Katherine J. Overstreet to Robert A. Leslie.
9 Jan McLaren, 7 August 2016; Nick Reid, 8 June 2017.
10 www.voyager.jpl.nasa.gov/mission/interstellar (accessed 8 June 2017); *Canberra Times*, 27 May 1979, 21 October 1980, 22 November 1985, 10 January 1986, 28 May 1988.
11 *Canberra Times*, 1 December 1981, 16 June 1982, 18 January 1986; Douglas J. Mudgway, *Uplink-Downlink: a History of the Deep Space Network 1957–1997*, The NASA History Series, NASA Office of External Relations, Washington DC, 2001, p. 197.
12 *Canberra Times*, 16 July 1989.

13 *Canberra Times*, 13 and 14 April 1981.
14 *Canberra Times*, 20 March 1983; Margaret Reid, 6 June 2017; Megan Reid's speech at Tom Reid's wake, October 2010.
15 *Canberra Times*, 28 May 1988; Telex dated 12 October 1988 from Tom Reid to NASA and to its worldwide tracking network; Glen Nagle interviewed in *The Guardian*, 14 August 2017; Bernie Scrivener, 'The Life and Times of a Glaswegian', 30 November 1988.
16 Margaret Reid, 6 June 2017; Danae Reid's speech; www.aph.gov.au/Senators_and_Members/Senators/The_President (accessed 13 June 2017); *Canberra Times*, 16 July 1989 and 22 March 1990; Senator Margaret Reid's Valedictory Speech, 6 February 2003.
17 Author's interview with David Doepel, 10 September 2017; Colin Mackellar, 19 September 2017.
18 *Canberra Times*, 5 August 1987; www.cdscc.nasa.gov/Pages/cdscc_history (accessed 3 June 2017); Luncheon program: Wentworth Hotel, 1 November 1969; Chaikin, pp. 161–62; Margaret Reid, 16 June 2017; Colin Mackellar, 19 September 2017.
19 Margaret Reid, 6 June 2017; Marg Reid's eulogy at Tom Reid's wake, October 2010; Simon Cullen, 1 July 2017.
20 Marg Reid's eulogy at Tom Reid's wake, October 2010.

Bibliography

Aeuckens, A., *The People's University*, South Australian Institute of Technology, Adelaide, 1978.

Armstrong, N., *A Life of Flight*, St. Martin's Press, New York, 2014.

Bate, P., 'Bernard Hague (1893–1960)', *The Galpin Society Journal*, Vol. 15 (March 1962).

Berry, S. and Whyte, H. (eds.), *Glasgow Observed*, John Donald Publishers, Edinburgh, 1987.

Blair, Anna, *Miss Cranston's Omnibus*, Lomond Books, Edinburgh, 1998.

Blair, Anna, *Tea at Miss Cranston's*, Shepheard-Walwyn, London, 1988.

Blaxland, J. and Crawley, R., *The Secret Cold War, The Official History of ASIO, 1975–1989*, Vol. III, Allen and Unwin, Sydney, 2016.

Catchpole, J., *Project Mercury*, Springer, Chichester, UK, 2001.

Chaikin, A., *A Man on the Moon*, Penguin, New York, 2007.

Clark, P., *Acquisition*, Elect Printing, Canberra, 2012.

Collins, M., *Carrying the Fire: an Astronaut's Journey*, Cooper Square Press, New York, 2001.

Dalzeil, N., *Glasgow*, The History Press, Stroud, Gloucestershire, 2009.

Dench, P. and Gregg., A., *Carnarvon and Apollo*, Rosenberg Publishing, Sydney, 2010.

Department of Defence, *An Outline of Australian Naval History*, Australian Government Publishing Service, Canberra, 1976.

Dougherty, K., *Space Australia, The Story of Australia's Involvement in Space*, Powerhouse Museum, Sydney, 2000.

Dudgeon, P., *Our Glasgow*, Headline Publishing, London, 2009.

Evans, S., *My Years as Director 1961–1978*, The South Australian Institute of Technology, Adelaide, 1978.

Faculty of Engineering, *Faculty Handbook*, University of Glasgow, n/d.

Faculty of Engineering, *Final Year Book 1948–1952*, University of Glasgow.

Firkins, P., *Of Nautilus and Eagles*, Cassel Australia, Sydney, 1975.

Fitzgerald, A., *Fitzgerald's Canberra*, Dalton Publishing, Canberra, 1969.

Frenkel, E., *Love and Math*, Basic Books, New York, 2013.

Goddard, D. and Milne, D. (eds.), *Parkes Thirty Years of Radio Astronomy*, CSIRO Publishing, Melbourne, 1994.
Hague, B., *An Introduction to Vector Analysis: for Physicists and Engineers*, Methuen, London, 1939.
Hancock, I., *He Did It His Way*, Hodder, Sydney, 2002.
Hetherington, H., (ed.), *Fortuna Domus*, University of Glasgow, Glasgow, 1952.
Houston, R. and Heflin, M., *Go, Flight! The Unsung Heroes of Mission Control, 1965–1992*, University of Nebraska Press, Lincoln, 2015.
Howse, D., *Radar at Sea, the Royal Navy in World War 2*, Naval Institute Press, Annapolis, Maryland, 1993.
Johnson, M., *Mission Control*, University Press of Florida, Gainsville, 2015.
Kranz, E., *Failure is not an Option*, Berkley Books, New York, 2001.
Lindsay, H., *Tracking Apollo to the Moon*, Springer, London, 2001.
McDougall, W., *The Heavens and the Earth a Political History of the Space Age*, Basic Books, New York, 1985.
Macleod, J., *River of Fire, The Clydebank Blitz*, Birlinn, Edinburgh, 20, C, 10.
McGuire, F., *The Royal Australian Navy*, Oxford University Press, Melbourne, 1948.
Maver, I., *Glasgow*, Edinburgh University Press, 2000.
Mayall, C., *Images of Scotland: Around Crieff and Strathearn*, The History Press, Stroud, Gloucestershire, 2009.
Morton, P., *Fire Across the Desert: Woomera and the Anglo-Australian Joint Project 1946–1980*, Department of Defence, Canberra, 1989.
Mudgway, D., *Uplink-Downlink: a History of the Deep Space Network 1957–1997*, The NASA History Series, NASA Office of External Relations, Washington DC, 2001.
National Capital Development Commission, *A Report on the Development of Canberra for the Five-Year Period July 1962 – June 1967*, Canberra, 1962.
Oakley, B., *A Mind for Numbers*, Penguin, New York, 2014.
Robertson, P., *John Bolton and a New Window on the Universe*, NewSouth Publishing, Sydney, 2017.
Shepard, A. and Slayton, D., *Moon Shot: The Inside Story of America's Race to the Moon*, Virgin Publishing, London, 1995.
Sizer, N., *The Road to Wealth*, Lee & Shepard, Boston, 1882.
Slayton, D. with Cassutt, M., *Deke!*, Forge, New York, 1994.
Southall, I., *Woomera*, Angus and Robertson, Sydney, 1962.
Sparrow, G., *Physics in Minutes*, Quercus, London, 2014.
Starchurski, R., *Below Tranquility Base*, CreateSpace Independent Publishing Platform, North Charleston, 2013.
Steven-Boniecki, S., *Live TV from the Moon*, Apogee Books, Burlington, Ontario, 2010.
Sullivan, B. and French, J., *To the Moon and Back*, Angus and Robertson, Sydney, 2004.
Swenson, L., Grimwood, J. and and Alexander, C., *This New Ocean: a History of Project Mercury*, Red and Black Publishers, St Petersburg, Florida, 1966.
Thom, A., *From the Days of the Horseless Carriage: Centenary of the Glasgow University Engineering Society*, Bell & Bain, Glasgow, 1991.
Tsiao, S., *Read You Loud and Clear; the Story of NASA's Spaceflight and Data Tracking Network*, The NASA History Series, Washington DC, 2008.

Verne, Jules, *From the Earth to the Moon and Around the Moon*, Wordsworth, London, 2011.
Walker, F., *Maralinga*, Hachette Australia, Sydney, 2014.
Ward, J., *Countdown to a Moon Launch: Preparing Apollo for Its Historic Journey*, Springer, 2015.
Ward, J, *Rocket Ranch: The Nuts and Bolts of the Apollo Moon Program at Kennedy Space Center*, Springer, New York, 2015.
Western Electric Company, *Introduction to Project Mercury and Site Handbook*, New York, 1960.
Wicks, B., *No Time to Say Goodbye*, Bloomsbury Publishing, London, 1988.
Wigmore, L, *The Long View: Australia's National Capital*, F.W. Cheshire, Canberra, 1963.
Williamson, J., *A History of Morrison's Academy*, Crieff, A.D. Garrie & Son, Auchterarder, 1980.
'Woomera...a Home on the Range', *Women's Weekly*, 3 December 1958.

Index